Food Microbiology Laboratory
for the Food Science Student

Cangliang Shen • Yifan Zhang

Food Microbiology Laboratory for the Food Science Student

A Practical Approach

 Springer

Cangliang Shen
Davis College, Division of Animal
    and Nutritional Sciences
West Virginia University
Morgantown, WV, USA

Yifan Zhang
Department of Nutrition and Food Science
Wayne State University
Detroit, MI, USA

ISBN 978-3-319-58370-9   ISBN 978-3-319-58371-6   (eBook)
DOI 10.1007/978-3-319-58371-6

Library of Congress Control Number: 2017943814

This Springer imprint is published by Springer Nature
The registered company is Springer International Publishing AG
The registered company address is: Gewerbestrasse 11, 6330 Cham, Switzerland

Preface

In order to help food science students understand the relationship between microorganisms and food products, it is important to develop a food microbiology laboratory textbook. Currently, a very limited amount of food microbiology laboratory manuals is commercially available, which makes it difficult for students to have a useful food microbiology laboratory manual with them. The aim of this book is to supply the food microbiology lab course instructor with a useful lab manual to teach this course and food science students and food microbiologists who will be working in the food industry with a readable and easily understood and followed laboratory manual. This book is designed to give students an understanding of the role of microorganisms in food processing and preservation; relation of microorganisms to food contamination, foodborne illness, and intoxication; general food processing and quality control; role of microorganisms in health promotion; and federal food-processing regulations. The listed laboratory exercises are aimed to provide a hands-on opportunity for the student to practice and observe the principles of food microbiology especially enumeration, isolation, and identification of microorganisms in foods. Students will familiarize themselves with techniques used to research, regulate, prevent, and control microorganisms in food and understand the function of beneficial microorganisms during the food manufacturing process. This book includes almost all pictures of each step of lab work, and all the pictures are coming from real lab practice and lab results. This lab manual is typically fit for small land-grant institutions that teach food microbiology lab classes and for non-land-grant universities who plan to develop a food science course curriculum.

Morgantown, WV, USA
Detroit, MI, USA

Cangliang Shen
Yifan Zhang

Acknowledgment

I appreciate my division director Dr. Robert Taylor who assigned me to develop and teach a food microbiology laboratory course when I joined West Virginia University. I got the inspiration of writing this food microbiology manual from my Ph.D. advisor Dr. John N. Sofos, university distinguished professor of Colorado State University. He taught me that "safety, quality, and quantity" are the three basic principles to conduct any food microbiology research almost 10 years ago when I was an amateur of food microbiology. I will pass this information to my future students and to the readers of this lab manual. I would love to greatly appreciate Ms. Lindsey Williams, communications manager at Davis College, West Virginia University, who has helped to take all pictures of all lab works and results in each chapter. I want to thank my graduate students Mr. KaWang Li and Ms. Lacey Lemonakis and my undergraduate teaching assistants Ms. Payton Southall, Jordan Garry, and Jessica Clegg for their generous help during the writing of this lab manual. The writing of Chaps. 10 and 11 obtained strong support from Mr. Mark Carlson at Western Kentucky University. Finally, I would love to thank my wife Dr. Yi Xu for her love and support. This lab manual is also dedicated to my newborn son Alex Shen.

Cangliang Shen
Division of Animal and Nutritional Sciences,
West Virginia University, Morgantown, WV

Contents

Authors

Cangliang Shen is an assistant professor/extension specialist at Davis College, Animal and Nutritional Science and Families and Health Center, at West Virginia University. Dr. Shen worked at IEH Laboratories and Consulting Group, USDA ARS, and Western Kentucky University as postdoctoral research microbiologist and assistant professor of microbiology after obtaining his Ph.D. from Colorado State University. Dr. Shen is currently teaching undergraduate-/graduate-level microbiology courses including general microbiology lecture with lab and food microbiology lab. Dr. Shen has 10 years of research experience in postharvest sanitizing procedures for reducing food safety risks on poultry meat products and fresh produce at both laboratory and pilot plant levels. Dr. Shen has published 23 papers in peer-reviewed food microbiology journals and has presented 25 times in national or international food science conferences.

Yifan Zhang is an associate professor in the Department of Nutrition and Food Science at Wayne State University, Michigan. She is an editorial board member of the *Journal of Food Protection* and a member of the International Association for Food Protection (IAFP), Institute of Food Technologists (IFT), and American Society for Microbiology (ASM). She earned her Ph.D. in Food Science from University of Maryland, College Park. Prior to joining the Wayne State University faculty, Dr. Zhang has worked as a postdoctoral researcher at The Ohio State University. Dr. Zhang's research is on the molecular characterization and risk analysis of foodborne bacteria, the food and agricultural reservoir of antimicrobial resistance, urban agriculture, and the development of novel food safety control. Dr. Zhang is currently teaching introductory food science and advanced food science at the undergraduate level and microbial food safety at the graduate level.

Abstract

We will provide food microbiology laboratory training for all students taking this class and emphasize the importance of lab notebook record, and finally we will show students the examples of good lab notebook record.

Keywords

Lab safety training • Notebook • Record

Three Principles of Conducting Food Microbiology Lab Work

Safety: Before doing anything in lab, ask yourself is it safe; otherwise do not do it.
Quality: All lab work should be based on scientifically high quality.
Quantity: Based on high quality, we will finish as much lab work as we can.

Introduction of Biosafety Level (BSL)

BSL-1: Microbial agents are not well known for causing disease and present minimal potential hazard to lab personnel, i.e., generic *Escherichia coli*, *Pseudomonas* spp.
BSL-2: Microbial agents cause moderate hazard to lab personnel and environment, i.e., *E.coli* O157:H7, *Salmonella* spp.
BSL-3: Microbial agents may cause serious or potentially lethal disease through the inhalation route of exposure, i.e., *Clostridium botulinum*.
BSL-4: Microbial agents pose a high individual risk of aerosol-transmitted laboratory infections and life-threatening disease, i.e., *Mycobacterium tuberculosis* (TB).

We only use *BSL-1 and BSL-2* microbial organism in this lab manual.

Lab Safety Rules

1. Always wash your hands with antimicrobial soap water before and after lab work.
2. Always wear a clean lab coat and gloves, and do not wear open toe shoes and very short pants when conducting lab work.

© Springer International Publishing AG 2017

C. Shen, Y. Zhang, *Food Microbiology Laboratory for the Food Science Student,*
DOI 10.1007/978-3-319-58371-6_1

3. Always wear eye protection (goggles) when staining or handling hazardous laboratory chemicals.
4. Clean and sanitize your working bench with a disinfection solution before and after lab work.
5. Report any lab accidents immediately.
6. Clearly label petri dishes, tubes, and flasks before lab work; label name, bench (seat) number, date, microorganism name, etc.
7. Place any biohazard trash into biohazard trash can.
8. Smoking, eating, drinking, or placing anything into your mouth is prohibited.
9. Do not handle any electronic devices (cell phone, ipad, laptop) while doing the lab work.
10. Please let your instructor know if you are pregnant or planning to be pregnant.
11. Know the location of nearest fire alarm pull station, fire extinguisher, eyewash station, and first aid kit.

Lab Notebook Requirement

Notebooks can be checked after each lab section before you leave the classroom.

The front cover of your notebook will be labeled with the following information: your name, course name, professor's name, and semester.

Reserve the first three pages of your notebook for a table of contents. You will update the table of contents each week.

Number all pages.

Each notebook entry for any given lab period will be formatted in the following way:

1. The Title.
2. A Purpose statement: Describe the purpose of the exercise. Ask yourself "What are the goals of this exercise?"
3. A Materials and Methods section: The Materials and Methods for each lab will be described at the beginning of the period by the instructor. You will have an opportunity to copy this information into your notebook at the beginning of each lab period. You should describe what was needed and the steps taken (including any modifications that were made). Be as specific as possible. Be sure to use *correct spelling* for all microorganism names and italicize scientific names.
4. A Results section: Record all observations in your lab notebook. Colored pencils/pens should be used to illustrate results (i.e., observations made with the microscope) – all figures/ tables must have a title and legend (a description of what is being shown – label all relevant information).
5. A Discussion: *Summarize* your findings, and *discuss* how the exercise helped you understand the learning objectives. *Describe* why something may not have worked and what you would do differently next time to improve the outcome.

Each section must be *labeled*.

Examples of Good Notebook Records (From Ms. Payton Southall)

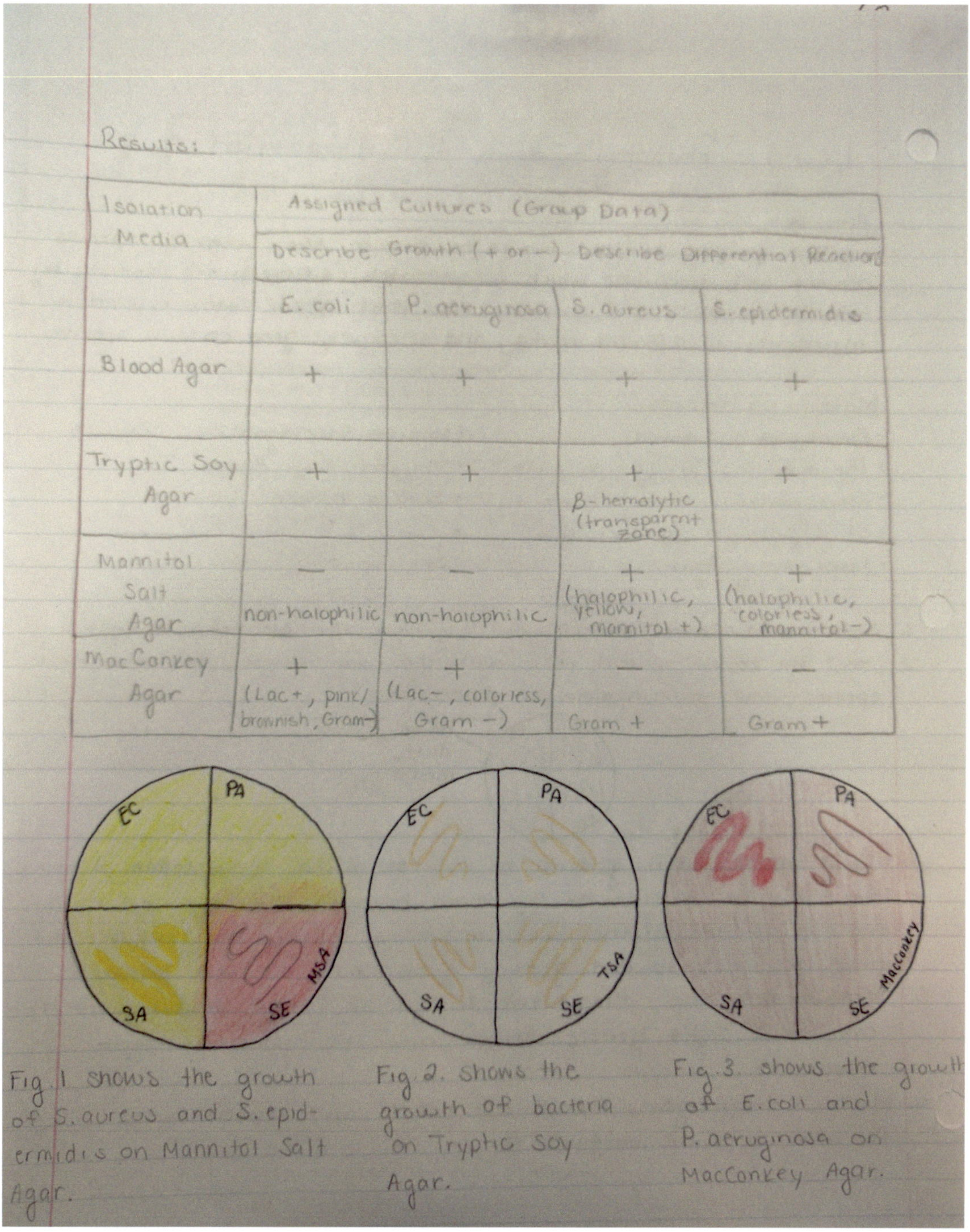

Results:

Isolation Media	Assigned Cultures (Group Data) Describe Growth (+ or −) Describe Differential Reaction			
	E. coli	P. aeruginosa	S. aureus	S. epidermidis
Blood Agar	+	+	+	+
Tryptic Soy Agar	+	+	+ β-hemolytic (transparent zone)	+
Mannitol Salt Agar	− non-halophilic	− non-halophilic	+ (halophilic, yellow, mannitol +)	+ (halophilic, colorless, mannitol −)
MacConkey Agar	+ (Lac+, pink/ brownish; Gram−)	+ (Lac−, colorless, Gram −)	− Gram +	− Gram +

Fig. 1 shows the growth of S. aureus and S. epidermidis on Mannitol Salt Agar.

Fig. 2. shows the growth of bacteria on Tryptic Soy Agar.

Fig. 3. shows the growth of E. coli and P. aeruginosa on MacConkey Agar.

Examples of Good Notebook Records (From Ms. Payton Southall)

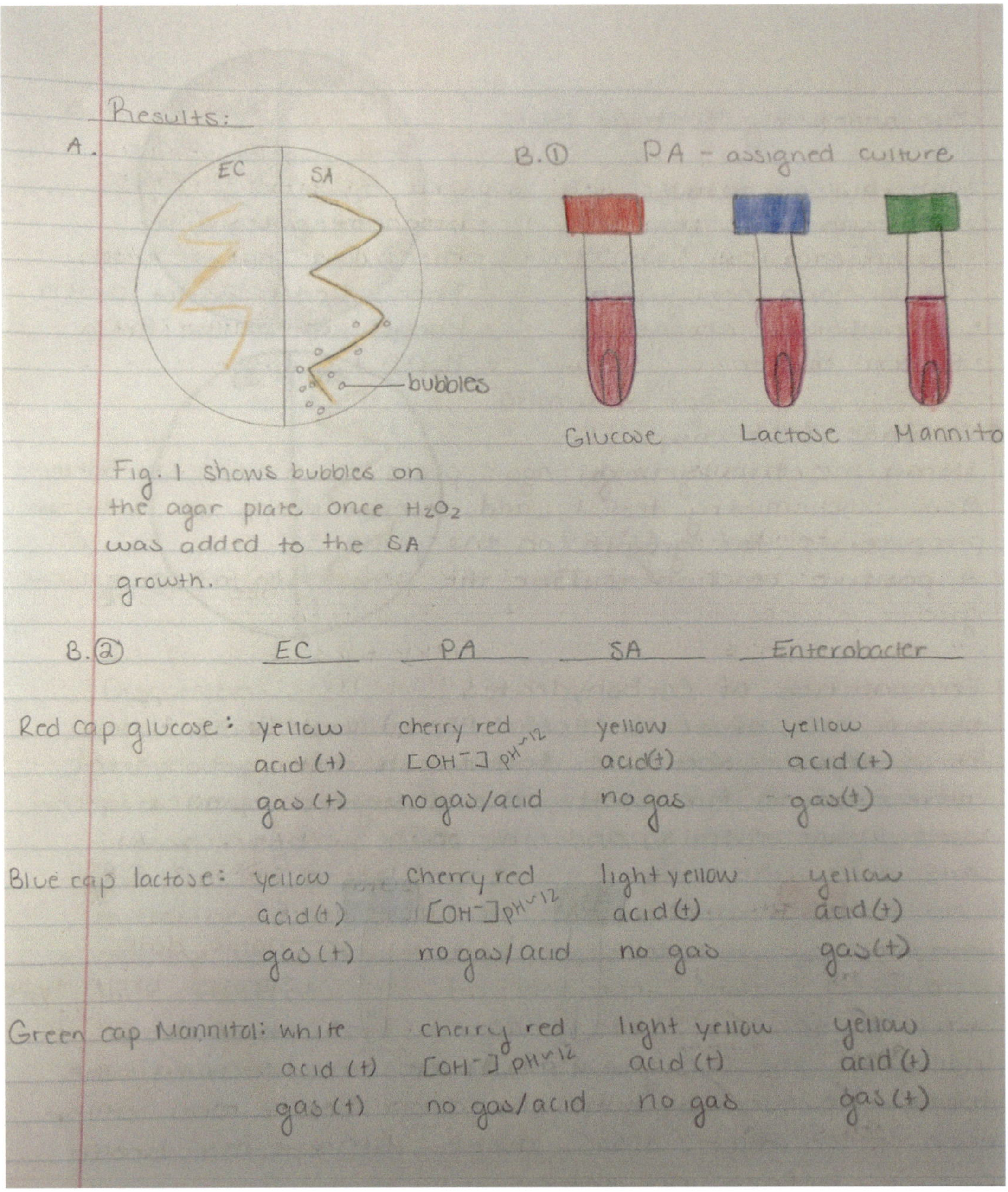

Results:

A.

EC | SA

—bubbles

Fig. 1 shows bubbles on the agar plate once H_2O_2 was added to the SA growth.

B.①   PA – assigned culture

Glucose   Lactose   Mannitol

B.②   EC   PA   SA   Enterobacter

Red cap glucose:
- EC: yellow, acid (+), gas (+)
- PA: cherry red, $[OH^-]\ pH \sim 12$, no gas/acid
- SA: yellow, acid (+), no gas
- Enterobacter: yellow, acid (+), gas (+)

Blue cap lactose:
- EC: yellow, acid (+), gas (+)
- PA: cherry red, $[OH^-]\ pH \sim 12$, no gas/acid
- SA: light yellow, acid (+), no gas
- Enterobacter: yellow, acid (+), gas (+)

Green cap Mannitol:
- EC: white, acid (+), gas (+)
- PA: cherry red, $[OH^-]\ pH \sim 12$, no gas/acid
- SA: light yellow, acid (+), no gas
- Enterobacter: yellow, acid (+), gas (+)

Examples of Good Notebook Records (From Ms. Jessica Clegg)

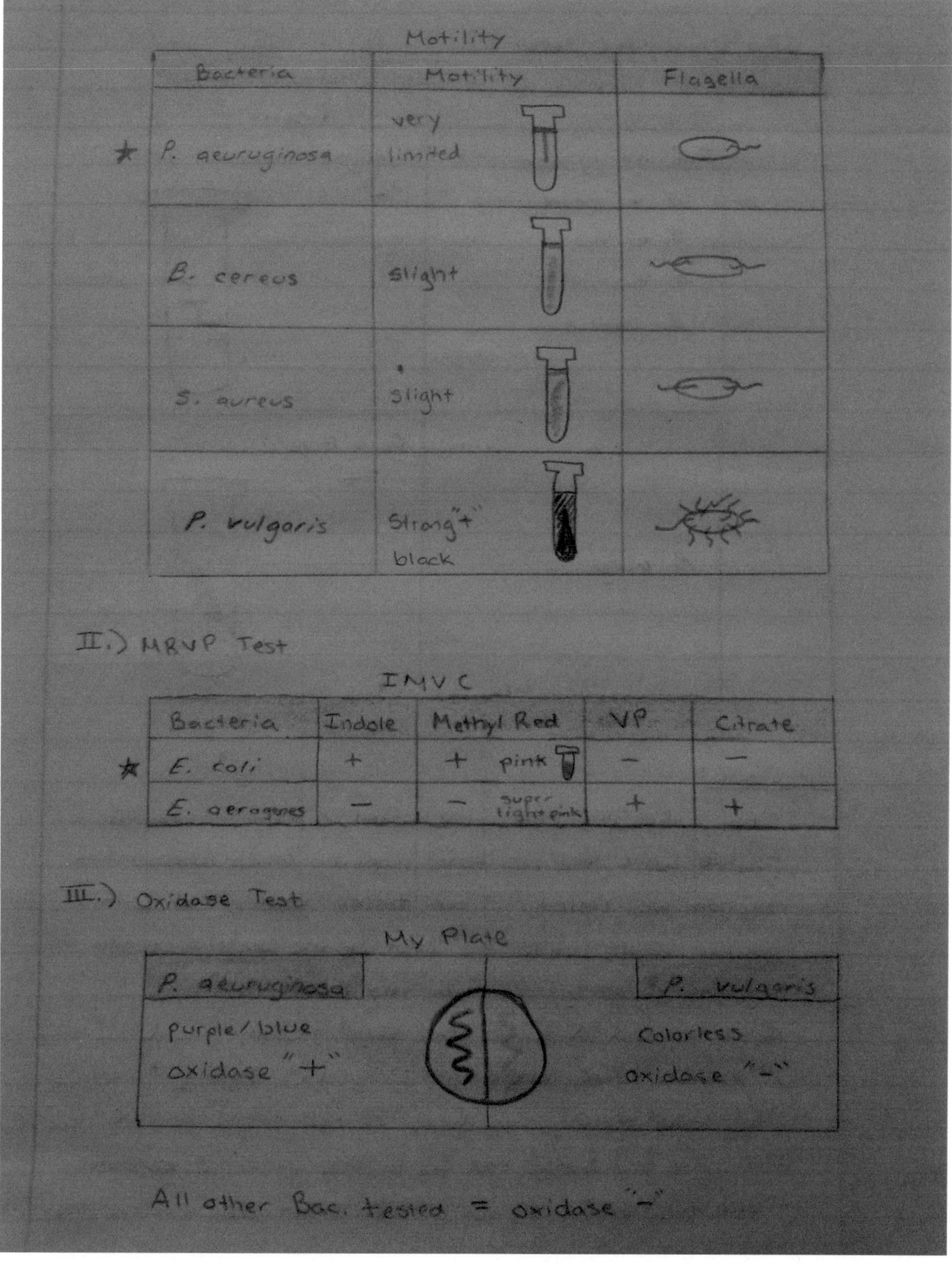

Examples of Good Notebook Records (From Ms. Jessica Clegg)

Results:

Isolation Media	Assigned Cultures (Group Data)		
	Describe Growth (+ or -) Describe Differential Reactions		
	E. coli	P. aeruginosa	S. aureus
Blood Agar	+	+	+ β-hemolytic
Tryptic Soy Agar	+	+	+
Mannitol Salt Agar	−	−	+ 7.5% Salt, halophile, yellow, mannitol +
MacConkey Agar	+ pink, Lac +	+ colorless, Lac −	− b/c Gram +

Discussion:

It was found that growth occured for all types of bacteria on
tryptic soy agar and blood agar. S. aureus growth on Blood agar
is β-hemolytic. Only S. aureus grows on mannitol salt agar and
is 7.5% salt, halophile, yellow, and mannitol +. E. coli and
P. aeruginosa grows on MacConkey Agar. S. aureus does not grow
on MacConkey b/c it is Gram +. E. coli on MacConkey agar is pink
and Lac + while P. aeruginosa is colorless and Lac −. This experiment
was helpful for recognizing importance of mannitol salt agar and MacConkey
agar in identifying bacteria.

Terms and Questions for Study

[Terms]

* Selective Growth Media - media that allows certain types of
   microorganisms to grow, and inhibits the growth of other
   microorganisms

* Differential Growth Media - media that is used to differentiate
   closely related microorganisms or groups of microorganisms
   owing to the presence of certain dyes or chemicals in the
   media

Class Notes

Abstract

We will introduce bright-field microscope use, practice Gram staining with foodborne pathogens, and practice endospore staining with *Bacillus cereus*.

Keyword

Bright-field microscope • Gram staining • Endospore staining • *Bacillus cereus*

Objective Practice Gram staining and endospore staining with common foodborne pathogens.

Gram Stain Strain Culture *Escherichia coli* ATCC25822, *Listeria monocytogenes, Salmonella enterica serovar* Typhimurium, *Staphylococcus aureus* ATCC 25923 (non-MRSA strain)

Endospore Stain Culture *Bacillus cereus*

Major Experimental Materials

- Bright-field light microscope
- Gram stain reagents (crystal violet, Gram iodine, 95% alcohol, safranin)
- Endospore reagent (malachite green)
- Sterilized loop
- Glass slides
- Lens paper
- Bunsen burner

Introduction of Gram Stain Gram stain was created by Hans Christian Gram (Danish microbiologist), which differentiates all bacteria into two groups *Gram positive (Blue/purple color)* and *Gram negative (Pink/red color)*. The theory of Gram stain differentiation is based on cell wall structure and lipid component; therefore, it comes out with two theories below:

C. Shen, Y. Zhang, *Food Microbiology Laboratory for the Food Science Student,*
DOI 10.1007/978-3-319-58371-6_2

1. Cell wall theory: Gram-positive bacteria have heavy peptidoglycan, which helps decolorizer (alcohol) to dehydrate the Gram-positive cell wall and traps the crystal violet complex inside the cell wall and maintains the purple color of crystal violet.
2. Lipids theory: Gram-negative bacteria have high lipid amount (10–15%) in the cell wall, which makes decolorizer (alcohol) easily to remove the crystal violet complex, and then colorless cell wall is stained with safranin and appears pink/red color.

Gram stains need to be done on a very young fresh culture (within 24–48 h).

Introduction of Bacterial Endospore Two bacteria genera that will generate endospore (germination) under stressful environment (poor nutrition, low humidity, desiccation, and high temperature) are *Bacillus* and *Clostridium*. The calcium-dipicolinic acid (DPA) stabilizes the endospore' DNA along with small soluble DNA protein to protect the bacteria from stress environment. When the growth condition improves, the endospore will generate vegetative cells (sporulation). Endospore stain is intended to find the presence/absence of endospore and the location of endospore. The location of endospore is species specific which can be located in the middle of cells (central) and the end (terminal, i.e., *Clostridium sporogenes*) or between the end and the middle (subterminal, i.e., *Clostridium botulism*). Endospore stain needs to be done on old culture (>5–7 days).

Gram Stain Procedure

1. Prepare a fixed smear: label two large circles (dime) using marker onto a clean glass slides, flip over slides, place one drop of sterilized water in the center of each circle, and use loop aseptically pick bacteria cells from stock culture and spread them in their drops of water to cover the whole circle of the dime (Fig. 2.1).

Fig. 2.1 Clean slides and label

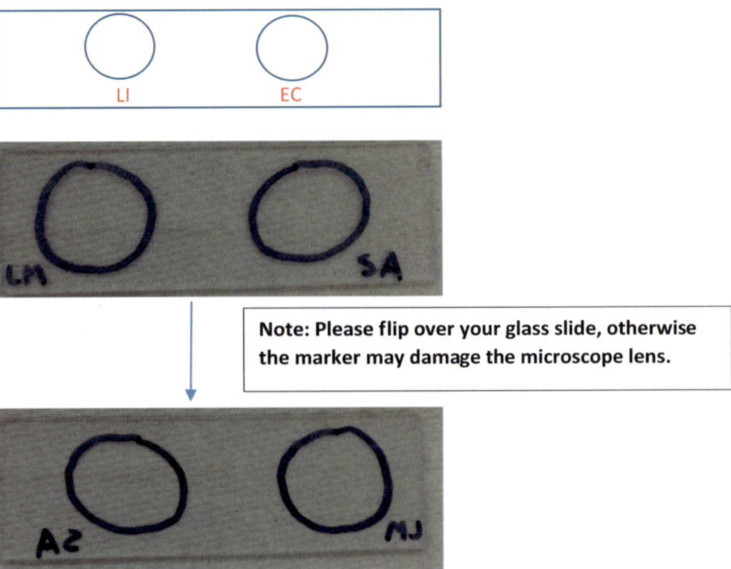

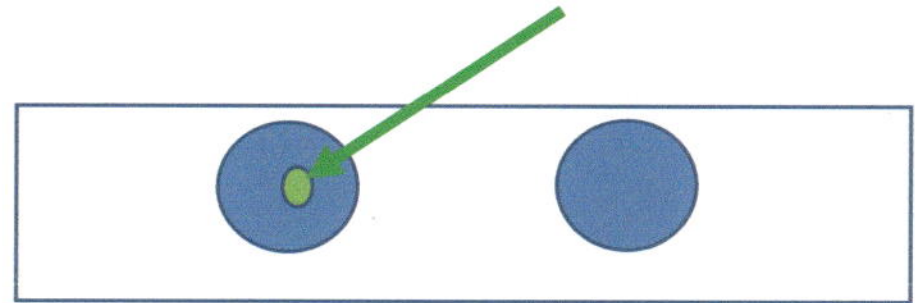

Fig. 2.2 Smear preparation, air-drying + heat fixing

Note: Flipping over slides is very important; otherwise, the dye from marker may damage the lens of microscope later on. A smear needs to be thin.

2. Air-dry the smear for 5 min (Fig. 2.2).
3. Heat fix the cells on slides by passing three times back and forth through the Bunsen flame. Heat fixing will inactivate bacteria enzyme and stick cells onto glass slides.
4. Major steps of Gram stain:
    (a) Place your slide on staining rack, and put one to two drops of *crystal violet* onto the smear for 60 s, and then gently rinse with water.
    (b) Add one to two drops of *Gram iodine* (a mordant to help crystal violet stain strong) for 60s, and then gently rinse with water.
    (c) Decolor by adding one to two drops of *95% alcohol* for 10–15 sec, and then gently rinse with water.
    (d) Counterstain by adding one to two drops of *safranin*, and then gently rinse with water.
5. Drain off excess water, blot dry with paper towel, and air-dry slide.
6. Observe your stained smear with low power 10X, and then switch to 100X with oil immersion; record your results in the notebook.
7. Gram staining results should include three items: (1) *Gram reaction*, positive or negative; (2) *morphology*, rods, cocci, etc.; and (3) *arrangement*, single, chain, grape shaped, tetrad, etc.
8. Please refer to Fig. 2.3 as a Gram staining result.

Endospore Stain Procedure

1. Smear preparation including air-drying and heat fixing is the same as Gram stain.
2. Flood the smear with *malachite green* (cover the whole slides with the dye, Fig. 2.4).
3. Heat steam slides for 3–5 min using the flame of burner back and forth a couple of times, but do not let the slides dry out.
4. Cool down slides, remove dye, and gently rinse with water.
5. Add *safranin* onto the slide.
6. Drain off excess water, blot dry with paper towel, and air-dry slide.
7. Observe your stained smear with low power 10X, and then switch to 100X with oil immersion; record your results in the notebook.

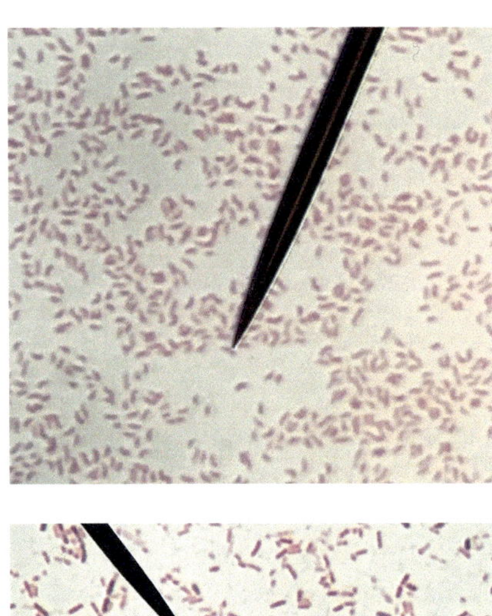

Escherichia coli

Gram stain negative, single, and short rods.

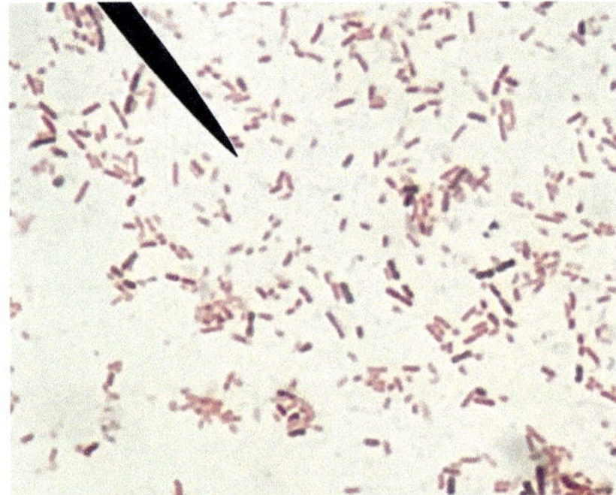

***Salmonella* Newport**

Gram stain negative, single, and short rods.

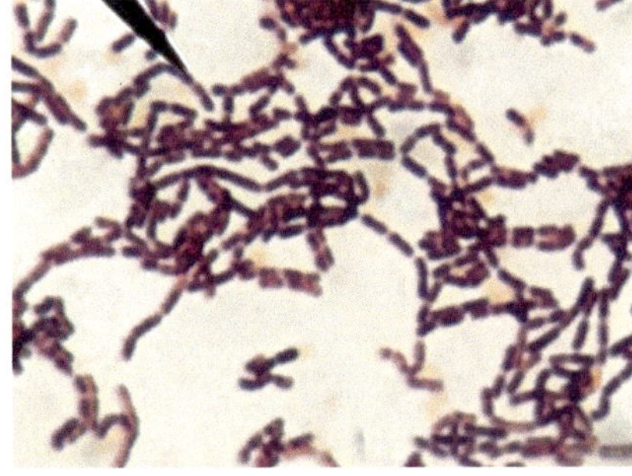

Listeria monocytogenes

Gram stain positive, single or chain, large rods.

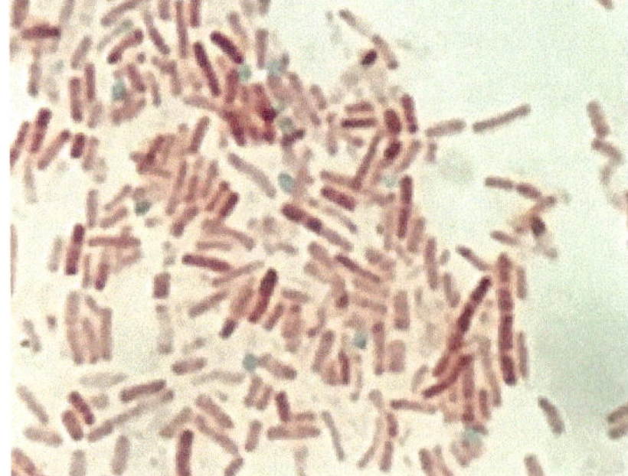

***Bacillus cereus* (48h old)** endospore stain large rods with terminal endospore (green color).

Fig. 2.3 Examples of Gram staining result (from previous students' lab work)

Fig. 2.4 Flooding with dye (only for endospore staining)

Bright-Field Microscope Use
Microscope structure

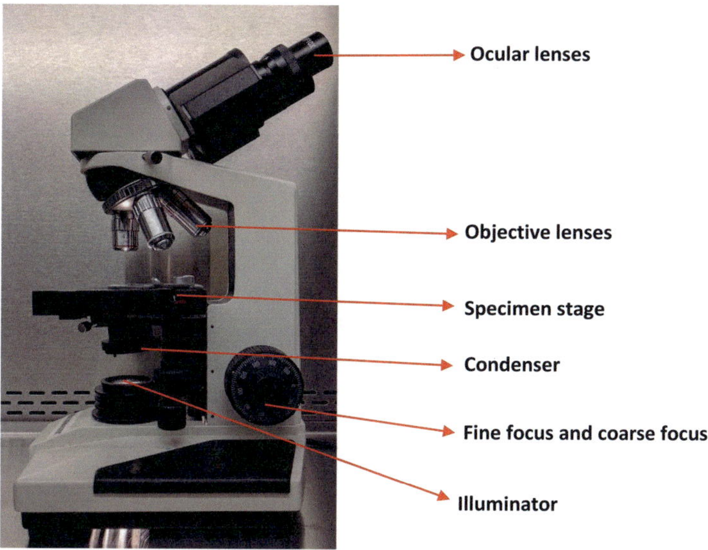

Brief Procedure

1. Stage down with course focus (large and thick wheel).
2. Spin/rotate objective lens.
3. Place 10X objective lens in position.
4. Adjust light to "low" or "10X" with condenser (iris diagram).
5. Bring stage up as high as it can reach.
6. Look into 10X ocular lens and find specimen.
7. Slightly switch 10X objective lens, and add one drop of oil. (Do not take microscope out of focus.)
8. Adjust fine focus (small and thin wheel) to find specimen.

Questions for Review

1. Why are Gram-positive stains purple and Gram-negative stains pink in color?
2. Describe two conditions in which bacteria organism will stain gram variable (showing Gram-negative results).
3. Is bacterial endosporulation a reproductive mechanism? Why?
4. Why is the location of endospore a very important information?

Class Notes

Enumeration of Bacteria in Broth Suspension by Spread and Pour Plating

3

Abstract

We will practice spread plating and pour plating using 10- or 100-fold dilution technique to enumerate bacteria cells in solutions on agars.

Keywords

Spread plating • Pour plating • Dilution technique • *Escherichia coli*

Objective Understand serial dilution technique, and practice spread and pour plating.

Bacterial Strain *Escherichia coli* ATCC25822 grown in Tryptic Soy Broth at 35 °C after 24 h.

Major Experimental Materials

- Tryptic Soy Agar
- Buffered peptone water (0.1%), 15 ml dilution tubes
- Water bath (50–55 °C) with melted agar (25 ml) in glass bottles
- Empty petri dishes
- Sterilized spreaders and pipette tips (200 μl and 1000 μl)

Introduction Most bacteria that grow in support medium like Tryptic Soy Broth incubating at 35 °C for 24 h will generate around 10^9 cells/ml ($=9.0$ $\log_{10}$ colony-forming unit/ml), which is too much to put onto agar medium to manually count the colony. Therefore, we need to do tenfold or 100-fold serial dilution and then add 0.1 ml of the diluted solution from the final dilution tube onto agar medium through spreading or pour plating and count bacterial colonies after incubation. For tenfold and 100-fold dilution, we usually use 9.0 or 9.9 ml 0.1% buffered peptone water (BPW). Spread plating is conducted by spread liquid solution on agar medium using sterilized plastic spreader or flamed glass spreader. Pour plating is conducted by adding liquid solution onto an empty petri dish followed by pouring 20–25 ml of melted agar, rotating the petri dish, and mixing very well. Due to the use of melted agar (50–55 °C), heat-sensitive bacteria will be killed by the mild heat; the number of colonies of pour plating will be slightly lower than spread plating.

© Springer International Publishing AG 2017
C. Shen, Y. Zhang, *Food Microbiology Laboratory for the Food Science Student*,
DOI 10.1007/978-3-319-58371-6_3

Dilution technique figures (tenfold serial dilution, assume 0.1 ml of each dilution tube adding onto agar plates)

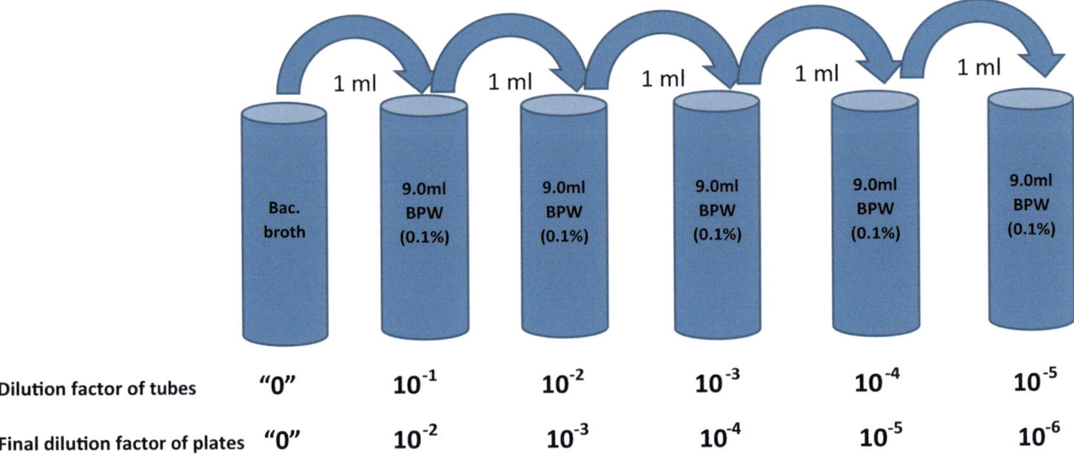

"0" Dilution Averagely spreading 1.0 ml original solution onto three agar plates, adding all colony-forming unit (CFU) of three agars after incubation.

Dilution Procedure

1. The original bacterial broth is considered to be at a dilution factor of "0."
2. Each dilution blank contains 9.0 ml of sterile 0.1% BPW.
3. To make a 1/10 dilution, transfer 1.0 ml of sample into a 9.0 ml sterile dilution blank (0.1%BPW). This is a 10^{-1} dilution.
4. Alternately, you can transfer 1.0 ml of diluent into a 9.0 ml blank and make a 10^{-2} dilution.
5. Continue to make dilutions as necessary.

Note: We also can add 0.1 ml into 9.9 ml of diluent to make a 10^{-2} dilution.
Dilution technique figures (100-fold serial dilution, assume 0.1 ml of each dilution tube adding onto agar plates).

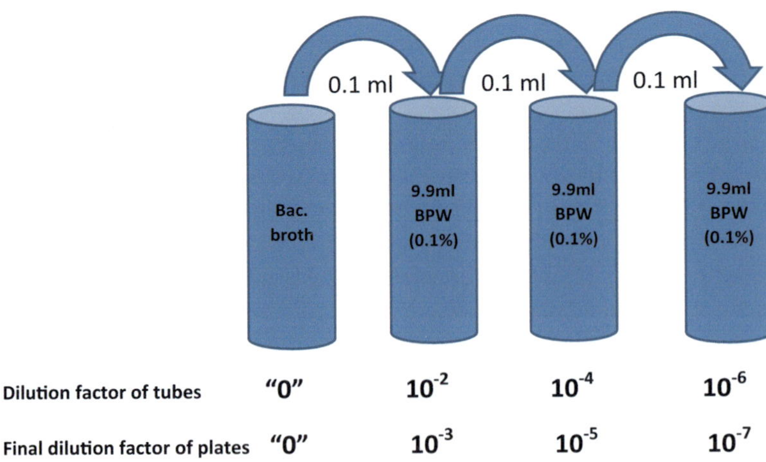

Numeration

After spread and pour plating, incubate plates inverted at 35 °C for 24–48 h, and manually count colonies on the agar plates. The acceptable and countable zone is 30–300 CFU/plate. CFU = colony-forming unit.

If colonies are <30, count and record it, but we will not use it, because colonies may come from contamination.

If colonies are 30–300, count and record it; the final counts will be (final colony number)/final dilution factor of plates, for example, 40 colonies on the plate with final dilution factor 10^{-6}, and then the final counts are $40/10^{-6} = 40{,}000{,}000$ CFU/ml = 7.60 $\log_{10}$CFU/ml.

If colonies on more than two plates are 30–300, count and record each and take average.

If colonies are >300, record as TNTC (too numerous to count).

Work figure of dilution procedure of plating ~ 10^9 CFU/ml E. coli solution onto agar plates.

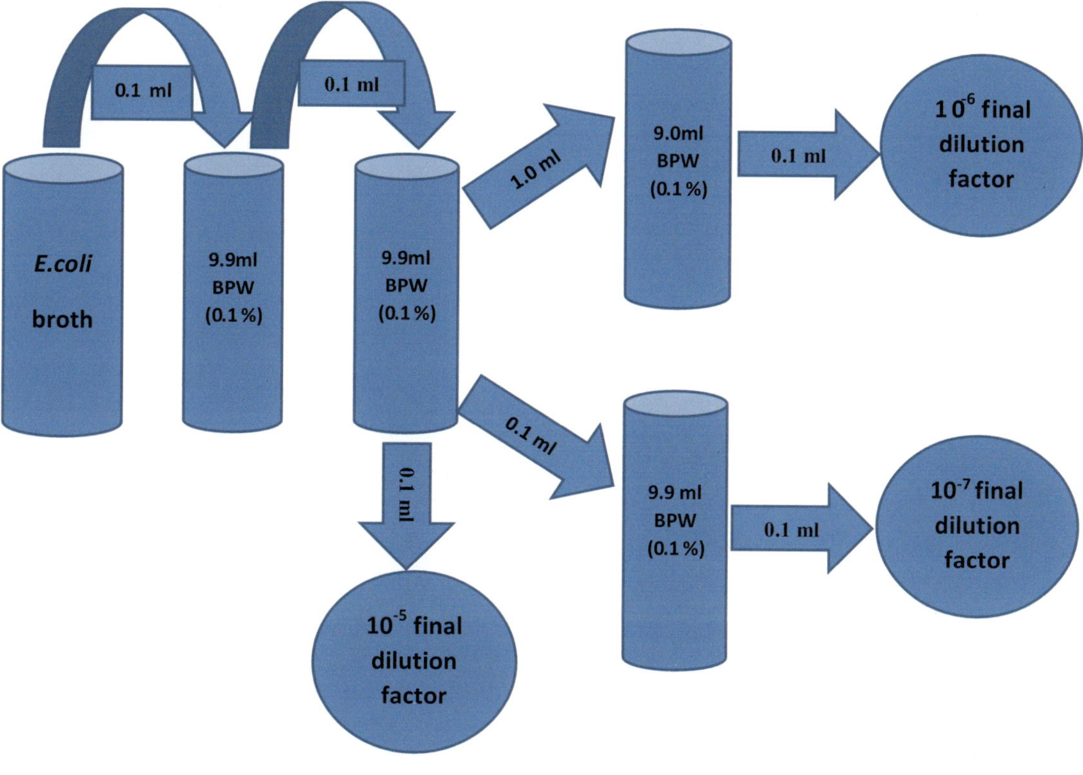

Method	10^{-5}	10^{-6}	10^{-7}	Reported count
Spread plating				
Pour plating				

Review Questions

1. A pure bacterial culture was diluted by adding a 0.1 mL aliquot to 9.9 mL water. Then, 0.1 mL of this dilution was plated out by spread plating, yielding 50 colonies. Calculate the CFU per mL in the original culture. ______________

2. A pure bacterial culture was diluted by adding a 1.0 mL aliquot to 9.0 mL water. Then, mL of this dilution was plated out by pour plating, yielding 82 colonies. Calculate the CFU/mL in the original culture. ______________

Class Notes

Isolation of Foodborne Pathogens on Selective, Differential, and Enriched Medium by Streak Plating

4

Abstract

We will introduce the selective, differential, and enriched medium for several foodborne pathogens, practice streak plating to get bacterial single colonies on various agars, and observe bacterial colony morphology.

Keywords

Streak plating • Single colony • Selective • Differentiate • Enrich • Blood agar • MacConkey agar • Mannitol salt agar • Modified Oxford Agar • XLT-4 agar • HardyCHROM agar

Objective Practice streak plating of foodborne pathogens on selective, differential, and enriched medium and observe colony morphology.

Bacterial Strain Generic *Escherichia coli* ATCC25822, *Listeria innocua* ATCC33090, *Salmonella enterica serovar Typhimurium*, and *Staphylococcus aureus* ATCC 25923 (non-MRSA strain).

Medium Blood agar, MacConkey agar, mannitol salt agar, Modified Oxford Agar, XLT-4 agar, and HardyCHROM agar.

Introduction of Selective and Differential Medium

Selective medium: Contains one or more components (chemical or antibiotics) that *suppress* the growth of some microorganisms without affecting the ability of wanted organism to grow.

Differential medium: Contains one or more chemical components which can *differentiate* different microorganism growth based on the *color and morphology* of colony.

Note: A medium can be both a selective and a differential medium.

Introduction of Medium for Isolating Foodborne Pathogens

MacConkey agar, containing neutral red, crystal violet, and bile salts, inhibits Gram-positive organisms (due to crystal violet) and supports Gram-negative organisms to grow. Colonies of *E. coli* O157:H7 are pink/red colony (due to lactose fermentation) surrounded by a bile salt precipitation zone. Non-lactose fermentation colony is colorless (e.g., non-O157 Shiga toxin-producing *E.coli*).

Mannitol salt agar, containing 7.5% salt, 6 carbon mannitol, and phenol red, is used for isolating *S. aureus* (yellow colony). The 7.5% salt inhibits most bacteria other than staphylococci. Mannitol can be fermented by *S. aureus* which cause phenol red to turn yellow color at acidic pH.

Modified Oxford Agar (MOX) contains oxford agar base plus selective ingredients, used for isolating *Listeria* spp. (black colony). Agar base provides basic nutrition of nitrogen, carbon, amino acids, and vitamins. *Listeria* spp. hydrolyze esculin to form 6,7-dihydroxycoumarin (esculetin), which reacts with ferric ions (ferric ammonium citrate) to form black colony. Lithium chloride inhibits growth of enterococci other than *Listeria* spp. (high salt tolerance). Selectivity is increased by adding colistin sulfate and moxalactam to the base and completely inhibits Gram-negative organisms and most Gram-positive organisms after 24 h of incubation.

XLT-4 agar is used for isolating non-typhi *Salmonella* (red colony with black center), which can ferment multiple sugar including xylose, lactose, sucrose, and decarboxylase of lysine and generate hydrogen sulfide. Hydrogen sulfide production (black color) is detected by the addition of ferric ions. Sodium thiosulfate is added as a source of inorganic sulfur. Sodium chloride maintains the osmotic balance of the medium. Phenol red is added as a pH indicator. XLT-4 supplement – Tergitol 4 – is added to inhibit growth of non-*Salmonella* organisms.

HardyCHROM Agar: HardyCHROM™ *Salmonella* agar facilitates the isolation and differentiation of *Salmonella* spp. from other members of the family *Enterobacteriaceae*. Peptones in the medium supply the necessary nutrients. *Selective agents* inhibit the growth of Gram-positive organisms. Artificial substrates (*chromogens*) are broken down by specific microbial enzymes which release insoluble colored compounds. *Salmonella* species break down only one of the chromogens and will produce deep pink- to magenta-colored colonies. Bacteria other than *Salmonella* spp. may break down the other chromogenic substrates and produce blue colonies. If none of the substrates are utilized, natural- or white-colored colonies will be present.

Blood agar, a Tryptic Soy Agar containing 5% sheep blood, differentiates pathogen growth based on the hemolysis of the red blood cells caused by hemolysin (exotoxin from *L. monocytogenes*).

Practice of Streak Plating

Streak plating is to "dilute" microorganism onto solid agar medium and then obtain pure culture (single colony represents one genera name and one species name) on agar surface:

Step 1. Use flamed loop to pick bacteria (agar, slant, or broth) and begin with inoculating from the first quadrant of the agar surface. Light touch without damaging the agar surfaces.
Step 2. Flame loop and cool down by touching the un-inoculated area of the agar surface.
Step 3. Rotate the plate, picking up area from the quadrant one, and streak again.
Step 4. Flame loop, rotate plate, and repeat procedure for quadrants three and four.
Step 5. Incubate plate inverted at incubator for 35 °C for 24 to 48 h.

Figure of Streak Plating

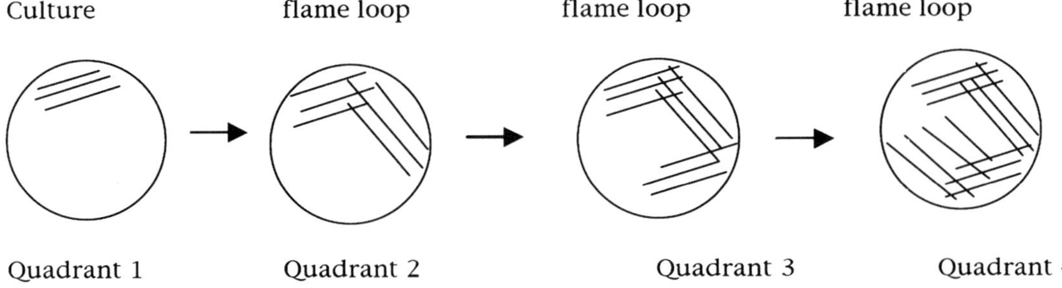

Lab Task Two students/group share seven agars and finish streak plating of assigned pathogen on a specific agar and "spot" plating with four pathogens on blood agar.

1. Streak-plating *Generic Escherichia coli ATCC25822* onto MacConkey agar, and incubate plates inverted at 35 °C for 24 h.
2. Streak-plating *Staphylococcus aureus ATCC 25923* onto mannitol salt agar, and incubate plates inverted at 35 °C for 48 h.
3. Streak-plating *Listeria innocua* ATCC33090 onto MOX agar, and incubate plates inverted at 35 °C for 48 h.
4. Streak-plating *Salmonella enterica serovar Typhimurium* onto XLT-4 agar and HardyCHROM agar, and incubate plates inverted at 35 °C for 24 h.
5. "Spot" plating of foodborne pathogens onto blood agar.

Divide a blood agar into four sections and label each section with one assigned culture.

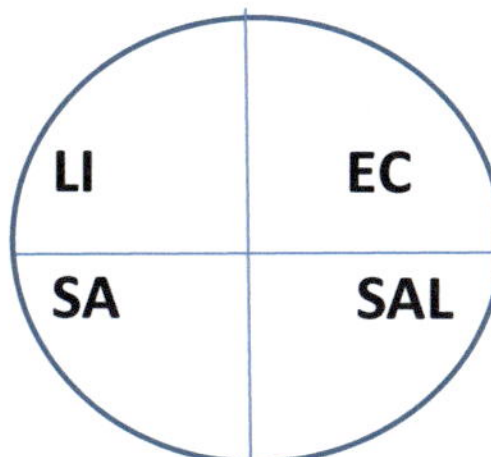

LI, L. innocua; EC, E.coli; SA, S. aureus; SAL, S. Typhimurium
Use your loop and "spot inoculate" each section with pathogen.
Incubate plates inverted at 35 °C for 24 h.

Question for Review

1. __________Mannitol salt agar (MSA) only allows the growth of halophiles (salt-loving microbes); nonhalophiles will not grow. Among the halophiles, mannitol fermenters will produce acid that turns the pH indicator yellow; mannitol nonfermenters leave the medium red. Onto MSA you inoculate a halophilic mannitol fermenter and a halophilic mannitol nonfermenter. In this case, the medium is acting as (a) __________ medium(s).
   A. Selective
   B. Differential
   C. Selective and differential
   D. Enriched
2. Briefly explain why *Salmonella* growing on XLT-4 agar is red colony with black center.
3. __________Phenol red is utilized in MacConkey agar to:
   A. Inhibit microscopic organisms
   B. Detect the production of gas by-products
   C. Conduct the sulfate reduction test
   D. Detect the acid reaction of lactose fermentation

Part of Result Figure (from Previous Lab Students)

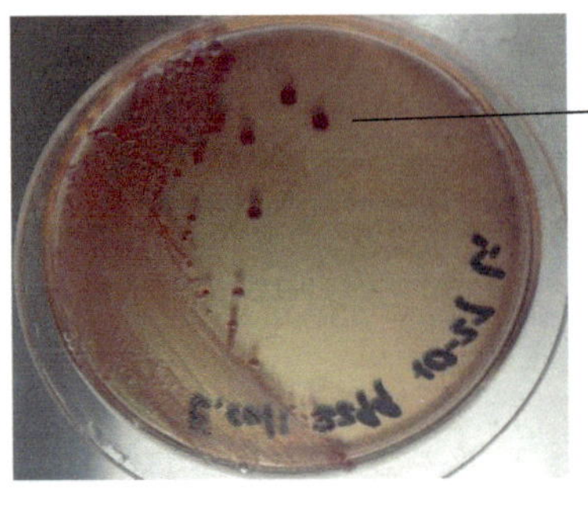

Escherichia coli on MacConkey agar: pick colony, lactose fermentation positive

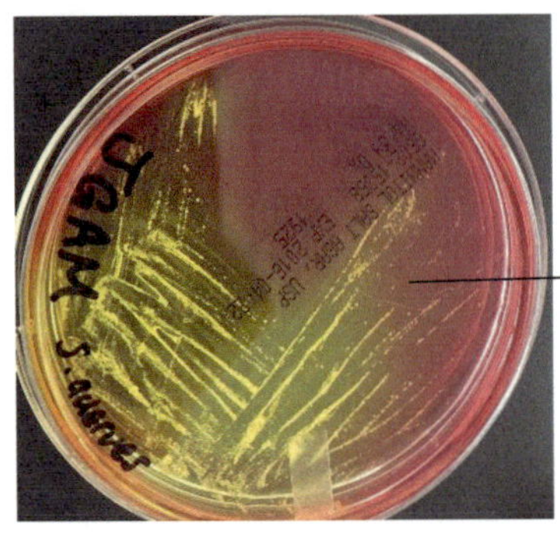

Staphylococcus aureus on Mannitol Salt agar: yellow colony, mannitol sugar fermentation positive (note: tiny colonies due to grow at 35°C for only 24h)

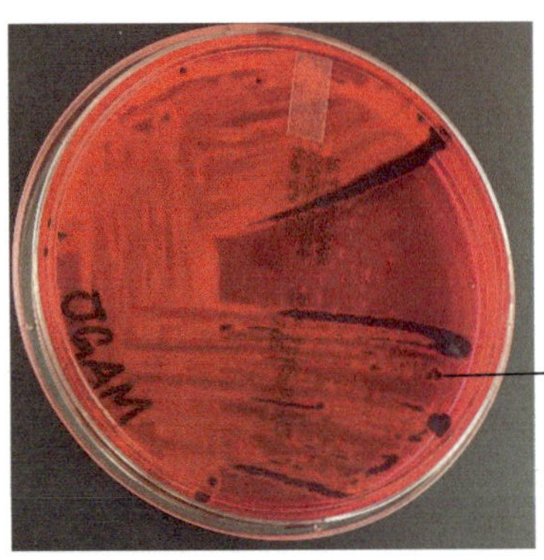

Salmonella spp. on XLT-4 agar: red colony with black center, suggests sulfur reduction (note: black color will be fade after 5 days due the oxidation of sulfur)

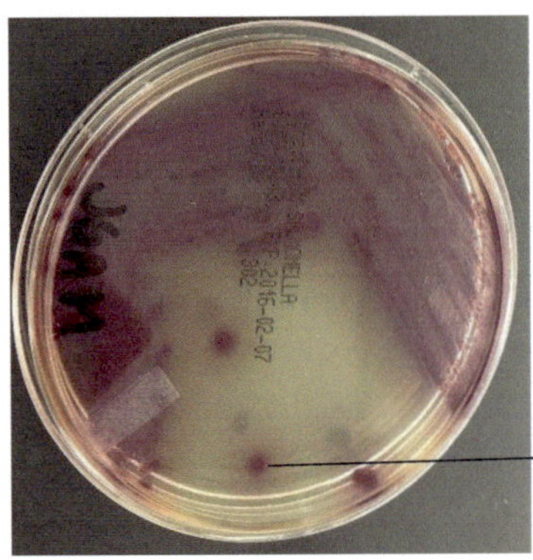

Salmonella spp. on Hardy®Chrom agar: deep pink colony

Class Notes

Enumeration of Aerobic Plate Counts, Coliforms, and *Escherichia coli* of Organic Fruit Juice on Petrifilm

Abstract

We will practice using Petrifilm to enumerate food microbial quality index including aerobic plate counts, coliform, and *E. coli* from organic fruit juice.

Keywords

Petrifilm • APCs • Coliform • *E. coli* • Juice

Objective Organic juice made from whole foods will be plating upon 3M® Petrifilm to test aerobic plate counts (APCs), *E. coli*, and total coliform populations.

Major Experimental Materials

- Organic fruit juice
- APC, ECC/TCC 3M® Petrifilm
- Buffered peptone water (0.1%), 15 ml dilution tubes
- Sterilized pipette tips (200 μl and 1000 μl)
- Incubator

Introduction

Indicator microorganism: Indicator organisms of foodborne pathogens are present in high number and easy to detect. Their growth and survival characteristics are similar to those pathogens.

Coliforms: They are most commonly used as indicator organisms, usually from the intestine of warm-blooded animals. They are Gram-negative, non-endospore-forming, facultative, fermentation lactose generating acid and gas and members of *Enterobacteriaceae* family.

E. coli: A member of the coliform group; the presence of *E. coli* may have more accurate correlation to the presence of pathogens than coliforms and indicates fecal contamination.

Petrifilm plates: Created by Food Safety Division of 3M Corporation, which contain a foam barrier accommodating the plating surface itself (a circular area of about 20 cm²) and a top film. Petrifilm has a cold-water-soluble gelling agent containing dehydrated nutrients and indicators for microorganism activity and enumeration.

Dilution factor APCs, 10^{-1} and 10^{-3}; coliform/*E. coli*, 0 and 10^{-2}; total coliform, 0, 10^{-1}, and 10^{-3}.

Dilution Protocol

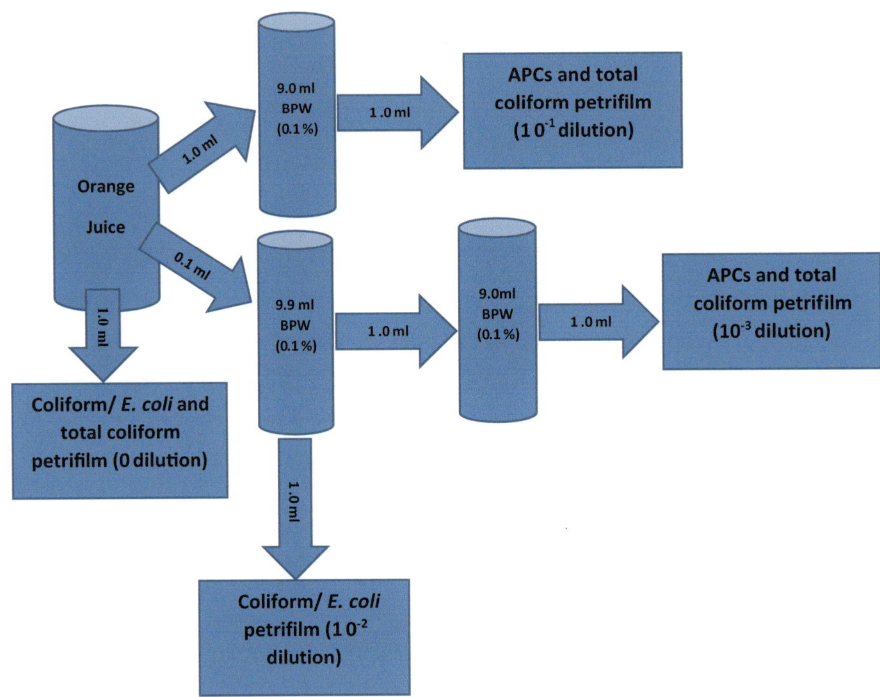

Procedure of Plating upon Petrifilm

1. Label each Petrifilm.
2. Make dilution according to the work protocol 0.1% BPW solution. Because 1 ml of solution will be plated onto Petrifilm, the final dilution is the same as tube dilution factor.
3. Pull up the film, and add 1 ml onto the center of Petrifilm with dehydrated medium.
4. Roll down the film and prevent air bubble.
5. Use plastic template to gently press down the sample to fill the area (coliform/*E. coli* film can skip this step).
6. Incubate at 35 °C for 24 to 48 h. Petrifilms cannot be stacked more than 20 pieces.

APCs, red colony; coliform, pink colony; *E. coli*, blue colony with gas.

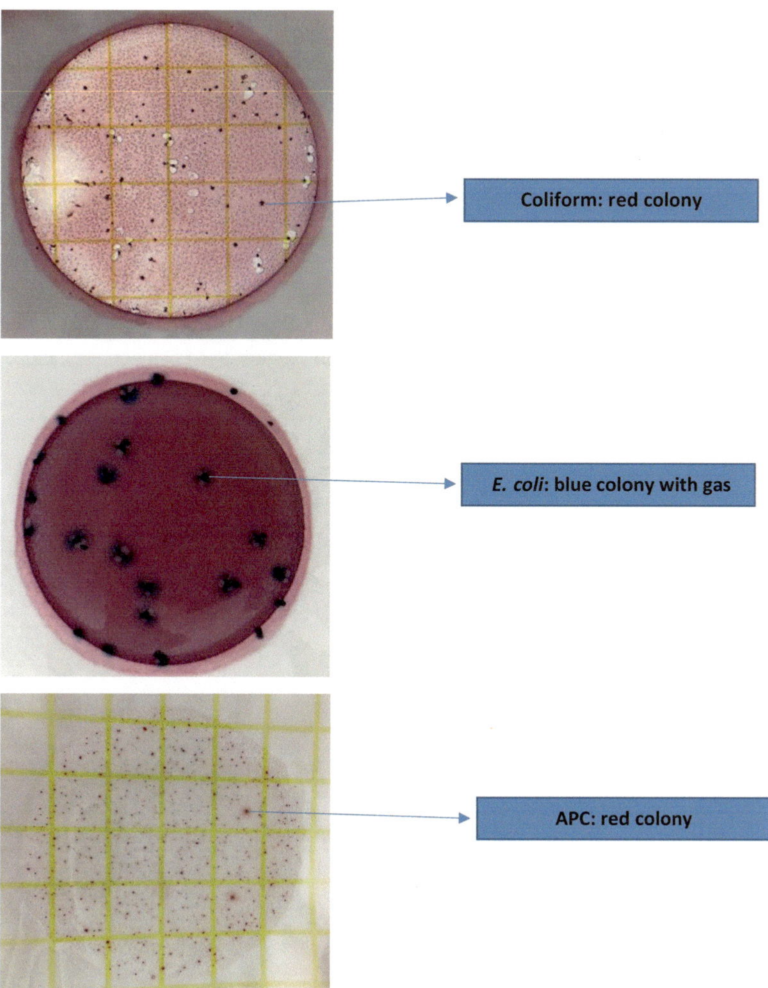

Review Question If you are a microbiologist working in a dairy farm and will test APCs and coliform/*E. coli* in unpasteurized raw milk, please write down your dilution flowchart in your notebook. The dilution factor will be APCs, 10^{-2} and 10^{-4}; coliform/*E. coli*, 0, 10^{-1} and 10^{-2}; and total coliform, 0, 10^{-1}, and 10^{-3}.

Lab work flowchart

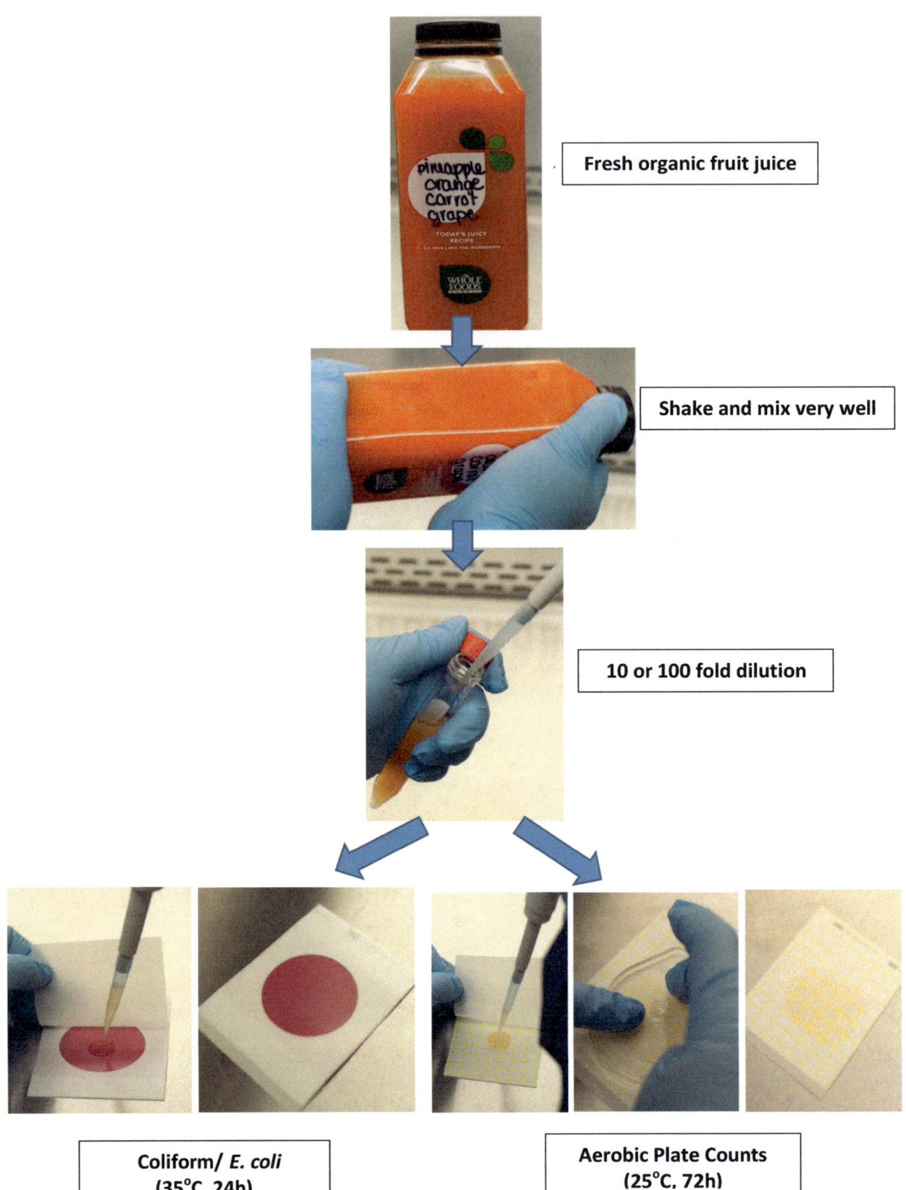

Part of Result Figures (Work from Previous Students)

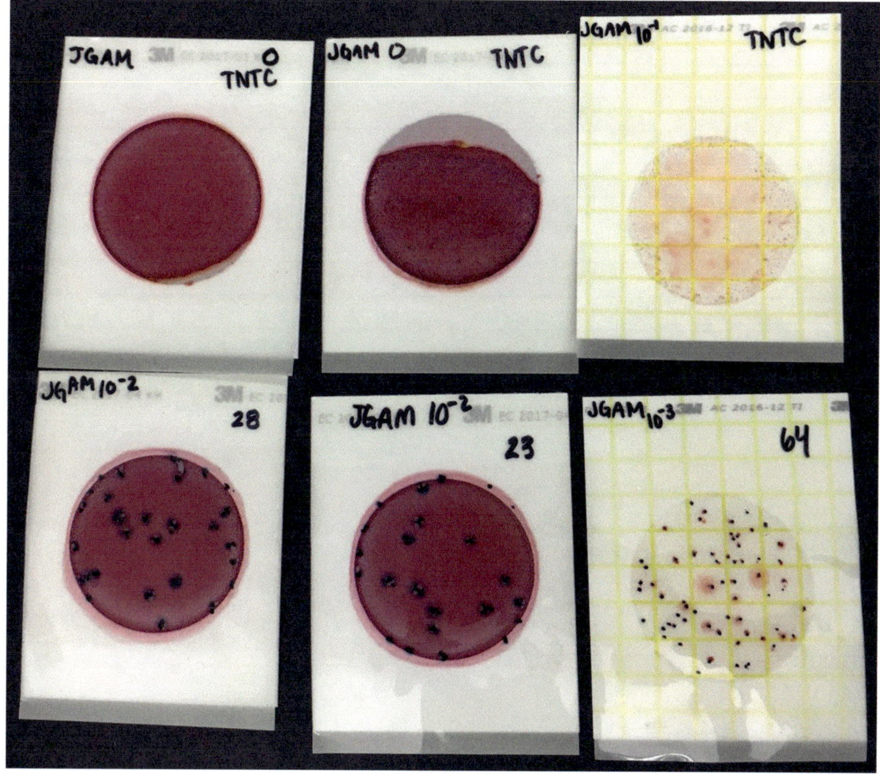

APC counts = $64 \times 10^3 = 6.4 \times 10^4$ CFU/ml = 4.81 $\log_{10}$ CFU/ml.

E. coli counts (less than 30 CFU on Petrifilm only estimated results) = $[(28 + 23)/2] \times 100 = 2550$ = 3.41 $\log_{10}$ CFU/ml.

Note: Coliforms are all *E. coli* cells in this lab result.

Class Notes

____
____
____
____
____
____
____
____
____
____
____
____
____

Enumeration and Identification of *Staphylococcus aureus* in Chicken Salads

6

Abstract

We will practice enumerating *Staphylococcus aureus* in chicken salads by inoculating, dilution technique, and spread plating onto selective medium. The presumptive colonies will be identified using Gram staining, catalase test, coagulase test, and latex agglutination test.

Keywords

Staphylococcus aureus • Chicken salads • Gram staining • Catalase test • Coagulase test • Latex agglutination test

Objective Test the population of *Staphylococcus aureus* in inoculated chicken salads using spread plating, and identify *S. aureus* colonies by Gram staining, catalase test, latex agglutination test, and coagulase test.

Major Experimental Materials

- *Staphylococcus aureus* ATCC 25923 (non-MRSA strain)
- Chicken salads from commercial supermarket
- Buffered peptone water (100% and 0.1%), 15-ml dilution tubes
- Sterilized spreader
- Sterilized pipette tips
- Whirl@ filtered food sampling bags
- Stomacher
- Mannitol salt agar, Tryptic Soy Agar
- H_2O_2 reagent (for catalase test)
- Gram stain reagents
- StaphTex™ Blue latex agglutination text kit
- Coagulase plasma (6 × 3 ml tubes)

© Springer International Publishing AG 2017

C. Shen, Y. Zhang, *Food Microbiology Laboratory for the Food Science Student,*

DOI 10.1007/978-3-319-58371-6_6

Introduction *Staphylococcus aureus* is the most commonly identified pathogen in all postsurgical infections with methicillin-resistant *S. aureus* (MRSA) becoming an emerging health problem. Staphylococcal food intoxication is a gastrointestinal illness caused by eating foods contaminated with *S. aureus*. Foods frequently incriminated include meat and poultry products, egg products, and salads such as egg, tuna, chicken, potato, and macaroni. Food handlers are usually the main source of food contamination in outbreaks. Symptoms develop within 6 h after eating contaminated foods, including nausea, vomiting, cramps, and diarrhea, and most often are self-limiting within 1 to 3 days.

Enumeration of *S. aureus* in Chicken Salads
Commercial chicken salads will be inoculated with *S. aureus* (your instructor will do this):

1. Inoculated chicken salads (200 g) will be sterilized transfer into filtered-whirl@ food sampling bag.
2. Add 200 ml of 0.1% buffered peptone water into the sample bag.
3. Stomach for 1 min.
4. Conduct serial dilution into 9.0 or 9.9 ml 0.1% BPW solution to obtain final dilution factor.
      10^{-2}, 10^{-3}, and 10^{-5} and spread plating onto mannitol salt agar and Tryptic Soy Agar. Please draw your dilution diagram and check with instructor!
5. Incubating plates at 35 °C for 48 h.
6. Manually count colonies of agar plates and record onto notebook, and calculate as follows:
      Log_{10} CFU/g = Log_{10} [(Final CFU on plates/dilution factor)*(200 g + 200 ml)/200 g]
      For example, 50 CFU on plates with dilution factor 10^{-5}, and then final Log_{10}CFU/g
      = Log_{10} $(50 \times 10^5 \times 2)$ = $\text{Log}_{10}10^7$ = 7 log_{10}CFU/g.
7. Typical colonies (yellow color) of *S. aureus* will be identified using the following tests:

A. *Gram stain*
   Major steps of Gram stain

   1. Place your slide on staining rack and put one to two drops of *crystal violet* onto the smear for 60 s, and then gently rinse with water.
   2. Add one to two drops of *Gram iodine* (a mordant to help crystal violet stain strong) for 60 s, and then gently rinse with water.
   3. Decolor by adding one to two drops of *95% alcohol* for 10–15 s, and then gently rinse with water.
   4. Counterstain by adding one to two drops of *safranin* and then gently rinse with water.

B. *Catalase test: positive (bubble)*
   1. Place one drop of H_2O_2 on a clean glass slides.
   2. Using sterilized loop, pick presumptive colony from mannitol salt agar and slowly immerse with cells into H_2O_2.
   3. A positive test is read immediately from the slide, showing oxygen gas bubbles.

C. *Coagulase test*
Transfer a presumptive colony from mannitol salt agar into the commercial coagulase plasma with EDTA and incubate at 35 °C and examine after 12 h for the clot formation. Firm and complete clot stays in place when tube is tilted or inverted.

D. *Latex agglutination test*

Latex agglutination test detects the agglutination (clumping) happened when the unknown organism isolated in a culture is mixed with an antiserum that contains antibodies specific for its antigen. The antibody (antiserum) usually is conjugated to a latex particle in order to enhance the visibility of the agglutination reaction.

Procedure

1. Place one drop of saline solution onto two wells of the test card.
2. Using sterilized loop, pick presumptive colony from mannitol salt agar and mix with the saline solution.
3. *Shake (very important!)* the latex reagent bottle (control and test reagent) and place one free-falling drop into each well.
4. Use a new sterilized loop to mix the reagent with colony very well.
5. Rock the card for 1 min to observe *the clumping*.

Review Question

A chicken salad was picked from a local subway and brought to the laboratory. 200 grams were added to 200 ml of sterile buffered peptone water and homogenized for 2 min. The solution was chilled to prevent overheating. The final dilutions and the collected data are presented below. The plate medium is TSA agar. Answer parts (a), (b), and (c).

(a) There are ________________ CFU per ml of homogenate.
(b) There are________________ CFU per gram of the chicken salad.
(c) What does CFU stand for?

____.

Plate	Colonies observed	Final dilution
A	Too numerous to count	10^{-1}
B	564	10^{-2}
C	275	10^{-3}
D	19	10^{-4}

Lab work flowchart

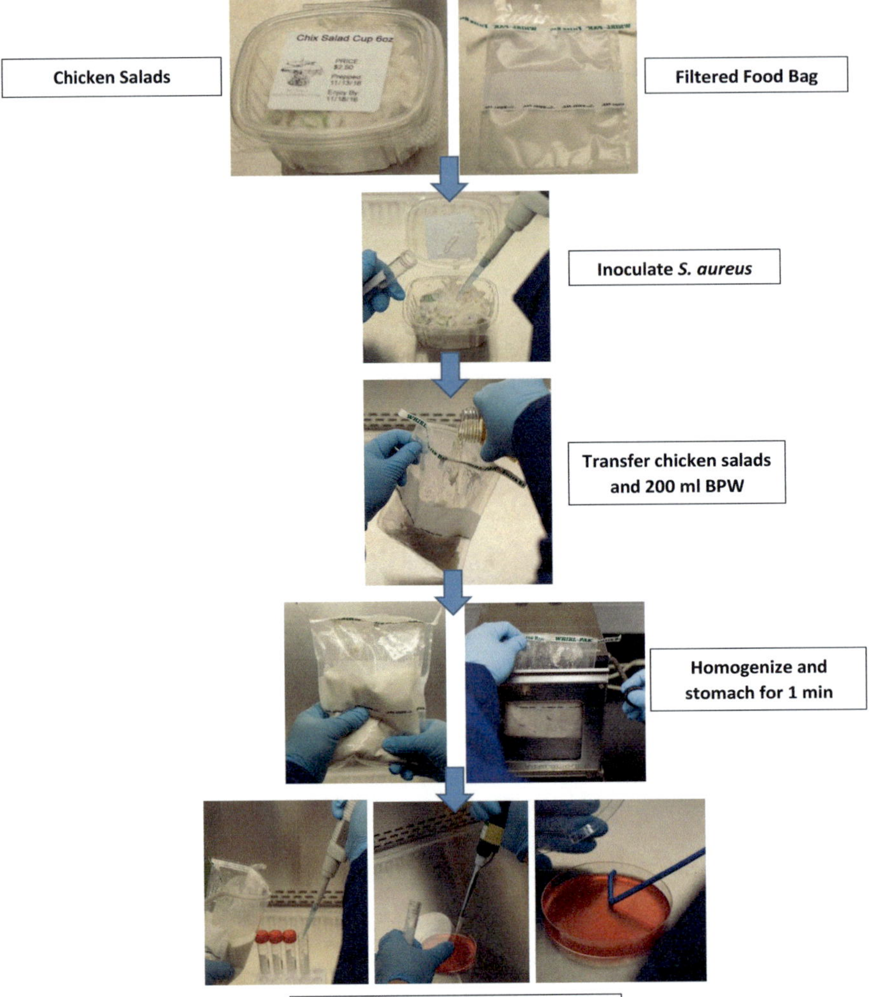

Part of Results Figures (Work from Previous Students)
A. Gram staining results (×100 with oil immersion)
Staphylococcus aureus is *Gram positive*, *cocci*, and *grape shaped*.

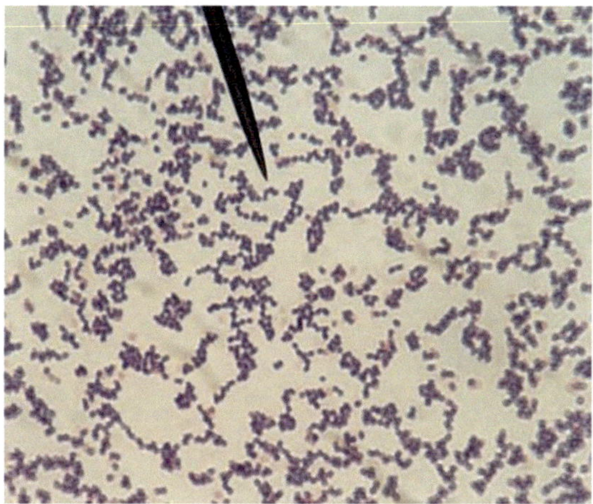

B. Latex agglutination test

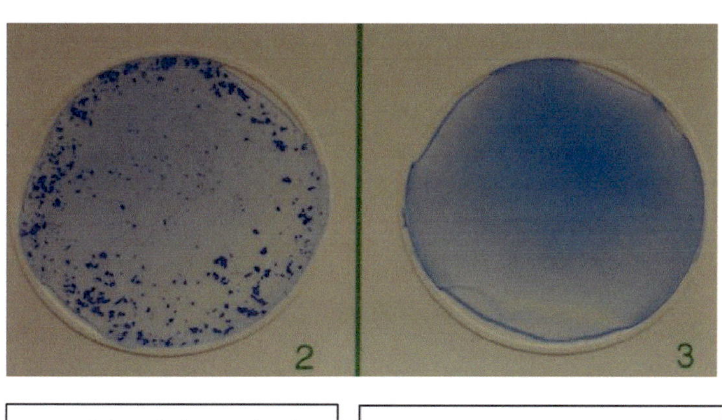

Positive (clumping)   **Negative**

C. Spead plating

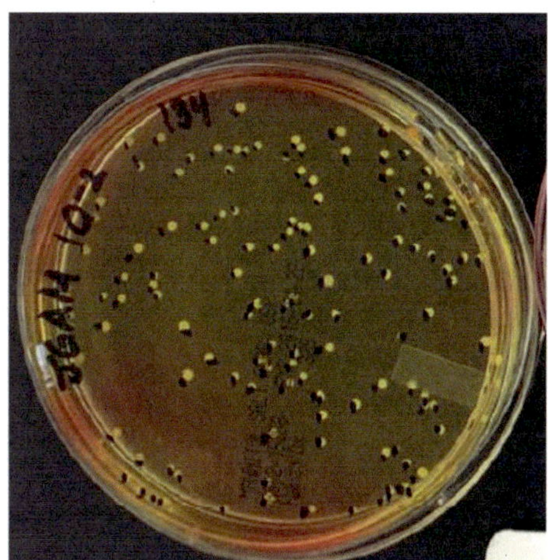

Yellow colonies suggest mannitol fermentation positive, total 134 CFU with dilution factor 100, final calculation:

134×100 × (200+200)/200=26800 CFU/g

Please note: this is not a perfect label, we should label date, bench number, and bacteria name.

Class Notes

Enumeration and Identification of *Listeria monocytogenes* on Ready-to-Eat (RTE) Frankfurters

Abstract

We will practice enumerating *Listeria monocytogenes* on commercial frankfurters by inoculating, dilution technique, and spread plating onto selective medium Modified Oxford Agar. The presumptive colonies will be identified using Gram staining, catalase test, and 12L test. In this chapter, it requires students to finish part of the work independently.

Keywords

Listeria monocytogenes • Frankfurters • Modified Oxford Agar • Gram staining • Catalase test • 12L test

Objective Test the population of *Listeria monocytogenes* on inoculated frankfurters by spread plating, and identify *L. monocytogenes* using Gram staining, catalase test, and 12L[@] biochemistry test.

Major Experimental Materials

- *Listeria monocytogenes* (BSL-2)
- Frankfurters (hotdog) from commercial supermarket
- Buffered peptone water (100% and 0.1%)
- Whirl[@] food sampling bags
- Stomacher
- Modified Oxford Agar, Tryptic Soy Agar, and blood agar
- H_2O_2 reagent (for catalase test)
- Gram stain reagents
- 12L[@] listeria test kit

Introduction *Listeria monocytogenes* is a Gram-positive, non-endospore-forming, facultative bacterium that can survive/grow in a wide variety of foods, including dairy products, meat and poultry products, vegetables, and seafood. *L. monocytogenes* is a remarkably difficult organism to control in food-processing environments due to its psychrotrophic nature and ability to tolerate adverse growth conditions. In the United States, from 1998 to 2002, ready-to-eat (RTE) meat products, including commercially cured ham, were involved in several multistate outbreaks of listeriosis. These outbreaks triggered the US Department of Agriculture's Food Safety and Inspection Service (USDA-FSIS) to require RTE meat processors, starting from 2003, to select at least one of three alternatives for *L. monocytogenes* control: (i) both a post-lethality treatment (e.g., steam, pressure, or antimicrobial agent) and an antimicrobial process, (ii) either a post-lethality system or an antimicrobial process, and (iii) sanitation practices only, with more frequent microbial verification testing.

Lab Work Procedure

Inoculation: Two frankfurter links will be received by two students/group and will be inoculated (0.2 mL) by rolling-spreading *L. monocytogenes* on the surface of each frankfurter (56 cm^2) with a sterile bent plastic rod on a foiled paper. The inoculated hotdogs were left at room temperature for 10 min to encourage attachment.

Enumeration of *L. monocytogenes* on Frankfurters

1. Inoculated two frankfurters will be transferred into whirl$^@$ food sampling bag.
2. Add 100 ml of 0.1% buffered peptone water into the sample bag and rolling bag.
3. Vigorously shake for 30 s (30 "Mississippi").
4. Conduct serial dilution into 9.0 or 9.9 ml 0.1% BPW solution to obtain final dilution factor.
   10^{-2} and 10^{-4} and spread plating onto Modified Oxford Agar and Tryptic Soy Agar. Please draw your dilution diagram in your notebook.
5. Incubating plates at 35 °C for 48 h.
6. Manually count colonies of agar plates and record onto notebook, and calculate as follows:
   Log_{10} CFU/g = Log_{10} [(Final CFU on plates /dilution factor)*100 ml)/112 cm^2]
   For example, 50 CFU on plates with dilution factor 10^{-4}, and then final Log_{10}CFU/cm^2
   = Log_{10} $(50 \times 10^4 \times 0.89)$ = Log_{10}445, 000 = 5.65 log_{10}CFU/g.
7. Typical colonies of *L. monocytogenes* will be identified using the following tests:

A. *Gram stain:* Gram positive, single, large rod shaped
   Major steps of Gram stain
   1. Place your slide on staining rack and put one to two drops of *crystal violet* onto the smear for 60 s, and then gently rinse with water.
   2. Add one to two drops of *Gram iodine* (a mordant to help crystal violet stain strong) for 60 s, and then gently rinse with water.
   3. Decolor by adding one to two drops of *95% alcohol* for 10–15 s, and then gently rinse with water.
   4. Counterstain by adding one to two drops of *safranin* and then gently rinse with water.
B. *Catalase test: positive (bubble)*
   1. Place one drop of H_2O_2 on a clean glass slides.
   2. Using sterilized loop, pick presumptive colony from MOX Agar and slowly immerse with cells into H_2O_2.
   3. A positive test is read immediately from the slide, showing oxygen gas bubbles.

C. *Hemolysis test:* Streak-plating typical colony onto blood agar, and observe α-, β-, or γ- hemolytic. *L. monocytogenes* and *L. innocua* are β-hemolytic (transparent zone).

D. *12L@ test*: Multiple channel biochemistry test for *Listeria* spp.; please follow instruction in the fact sheet.

Procedure

1. Pick four to five suspect colonies from MOX agar and suspend in *Listeria* suspending medium.
2. Place test strip in holding tray and remove lid.
3. Place four drops (100 µl) of bacterial suspension to each well.
4. Add one drop hemolysis to well 12. (We will do this on blood agar.)
5. Replace lid and incubate at 35 °C ± 2 °C for 24 h.
6. Record results on report forms and interpret using the color sheet.

12L positive and negative results table

Well no.	Substrate	Negative reaction	Positive reaction
1	Esculin (ESC)	Pink or brown	Black
2	Mannitol (MAN)	Purple	Yellow, brown, or straw
3	Xylose (XYL)	Purple	Yellow, brown, or straw
4	Arabitol (ARL)	Purple	Yellow, brown, or straw
5	Ribose (RIB)	Purple	Yellow, brown, or straw
6	Rhamnose (RHA)	Purple	Yellow, brown, or straw
7	Trehalose (TRE)	Purple	Yellow, brown, or straw
8	Tagatose (TAG)	Purple	Yellow, brown, or straw
9	Glucose-1-phosphate (G1P)	Purple	Yellow, brown, or straw
10	Methyl-D-glucose (MDG)	Purple	Yellow, brown, or straw
11	Methyl-D-mannose (MDM)	Purple	Yellow, brown, or straw
12	Hemolysin (HEM)	Red cell deposit	Partial or complete red cell lysis

Note: well 12 can be done on the blood agar.

Listeria spp. should be positive in well 1 (esculin), well 4 (arabitol), and well 7 (trehalose).

	1ESC	2MAN	3XYL	4ARL	5RIB	6RHA	7TRE	8TAG	9G1P	10MDG	11MDM	12HEM
L. monocytogenes	+	−	−	+	−	+	+	−	−	+	+	+

Review Questions

1. Name the "three alternatives" for controlling *Listeria monocytogenes* on RTE meats.
2. There is a *Listeria monocytogenes* outbreak in a hotdog plant; as a microbiologist, your job is to isolate and identify this pathogen from hotdogs and describe the details how you will do this, and please be specific, including medium, test name, and possible results.

Lab work flowchart

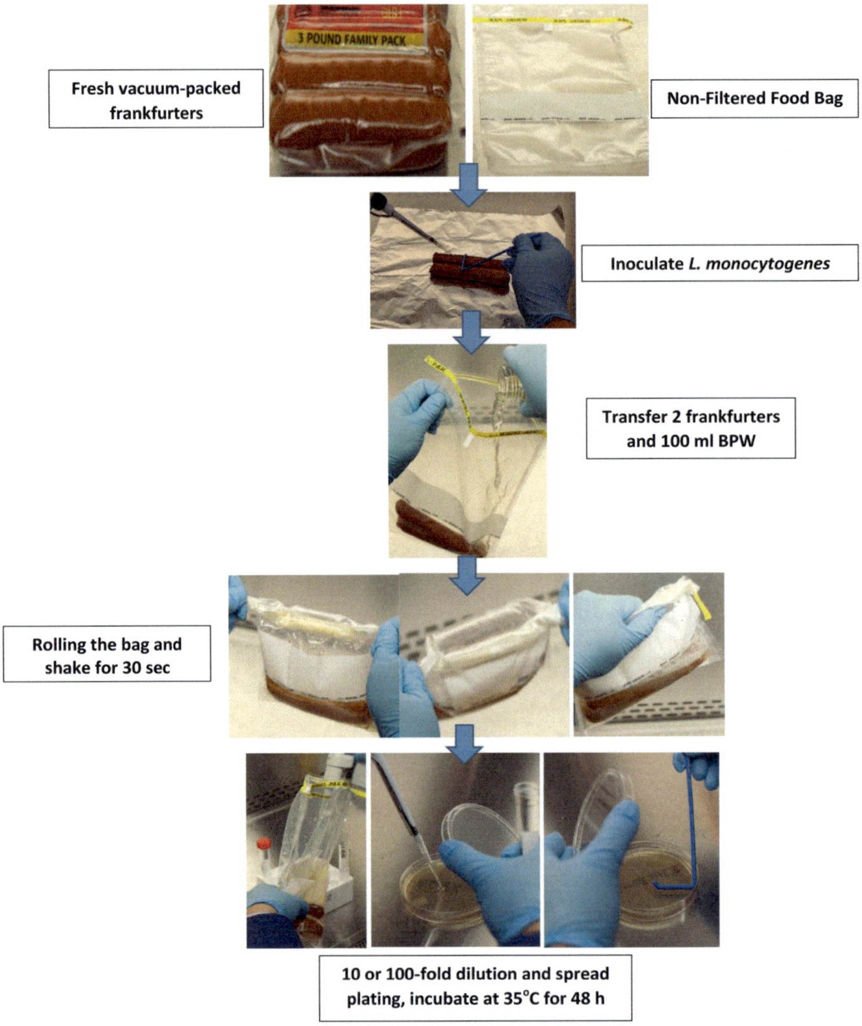

3 POUND FAMILY PACK
Fresh vacuum-packed frankfurters
Non-Filtered Food Bag
Inoculate *L. monocytogenes*
Transfer 2 frankfurters and 100 ml BPW
Rolling the bag and shake for 30 sec
10 or 100-fold dilution and spread plating, incubate at 35°C for 48 h

Part of Result Figures (Work from Previous Students)

A. Positive L. monocytogenes by 12L test

OXOID
MB12L
ESC
MAN
XYL
ARL
RIB
RHA
TRE
TAG
G1P
MDG
MDM
HEM
05689

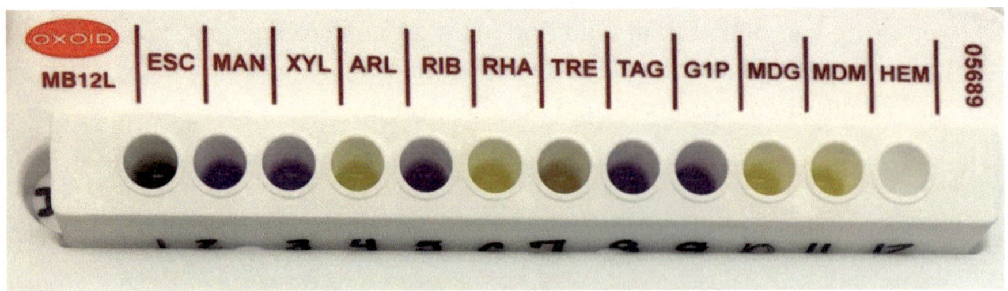

B. β-Hemolytic (transparent zone) of L. monocytogenes on blood agar

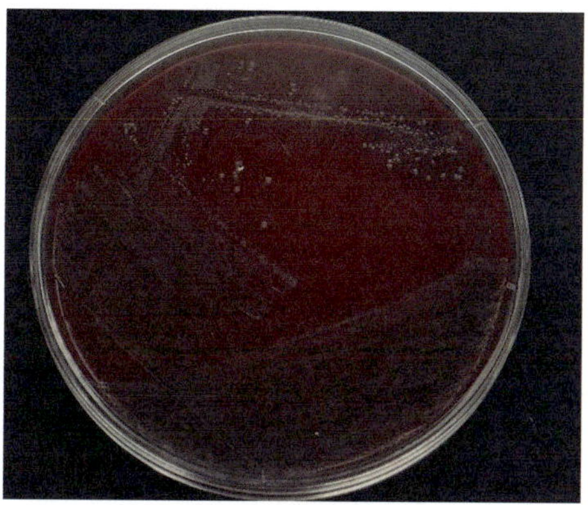

C. Spread plating

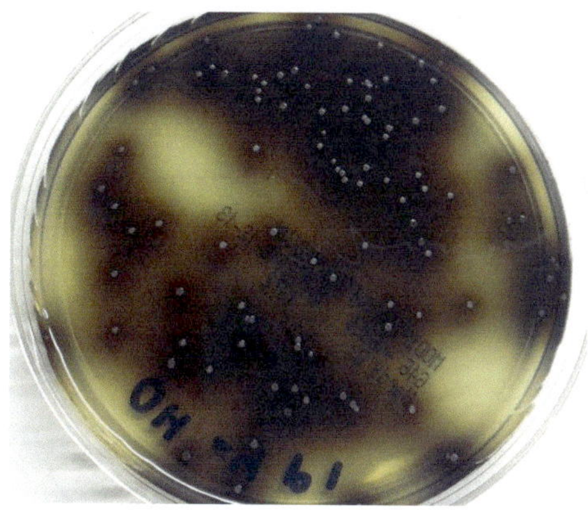

Class Notes

Isolation and Identification of *Salmonella* and *Campylobacter* spp. on Broiler Carcasses

Abstract

We will practice isolating and identifying *Salmonella* and *Campylobacter* spp. on commercial chicken broiler carcasses by modified USDA methods. We will introduce the function of primary and secondary enrichment and practice growing *Campylobacter* in a microaerophilic environment. The presumptive colonies will be identified using Gram staining, latex agglutination test, and API 20e biochemistry test. In this chapter, it requires students to finish part of the work independently.

Keywords

Salmonella • *Campylobacter* • Broilers • RV • TT • XLT-4 • HardyCHROM • Modified Campy-Cefex agar • Bolton broth • Gram staining • Latex agglutination test • API 20e test

Objective *Salmonella* and *Campylobacter* spp. will be isolated from commercial broiler carcasses and identified using Gram staining, biochemistry test, and immunology test.

Major Experimental Materials

- Raw chicken carcass
- Poultry sampling bag
- Buffered peptone water
- RV medium and TT medium
- XLT-4 and HardyCHROM agar
- Bolton broth and modified Campy-Cefex agar
- Microaerophilic jar and gas generator
- Campy-latex agglutination test kits
- API 20E biochemistry test kits
- Mineral oil
- Kovac reagent, 10% ferric chloride, 40% KOH, and 6% alpha-naphthol
- Gram stain reagents

Introduction Contaminated poultry meat represents the greatest public health impact among foods and is responsible for an estimated \$2.4 billion in annual disease burden. *Salmonella* and *Campylobacter* spp. are the two most common foodborne pathogens associated with poultry meat, causing an estimated 9.4 million illnesses, 55,961 hospitalizations, and 1351 deaths annually in the United States. Starting in July 2011, USDA-FSIS established new performance standards in response to national baseline studies requiring routine testing for *Salmonella* and *Campylobacter* in all processing plants, where the percentage of *Salmonella*-positive samples must be below 7.5% and *Campylobacter*-positive samples should be less than 10.4%.

Major Medium Used for *Salmonella* Isolation

Rappaport Vassiliadis (RV) Salmonella Enrichment Broth: Malachite green is inhibitory to organisms other than *Salmonella* spp. The *low pH (5.2)* of the medium, combined with the presence of malachite green and *magnesium chloride*, selects for the highly resistant *Salmonella* spp.

Tetrathionate (TT) Broth: Selectivity is accomplished by the combination of *sodium thiosulfate* and *tetrathionate*, which suppresses commensal intestinal organisms. Tetrathionate is formed in the medium upon addition of *the iodine and potassium iodide solution.* Organisms containing the enzyme tetrathionate reductase will proliferate in the medium. *Bile salt*, a selective agent, suppresses coliform bacteria and inhibits Gram-positive organisms. Calcium carbonate neutralizes and absorbs toxic metabolites.

Major Medium Used for *Campylobacter* Isolation

Modified Campy-Cefex Agar: A *Brucella* agar (agar base plus horse blood and hemin) supplemented with cephalothin and amphotericin B for the selective isolation of cephalothin-resistant *Campylobacter* species such as *C. jejuni, C. coli,* and *C. lari* from food samples.

Lab Work Procedure

Today

1. Each group (two students) will pick up one broiler carcass.
2. The broiler carcass will be immersed into 400 ml buffered peptone water (BPW) in a poultry sampling bag, and vigorously shake for 60 s.

Please do this within a large container; make sure the bags are not broken!

3. Transfer 25 ml of shaken BPW solution into 25 ml of Bolton broth in a tube, and mix very well. These mixtures will be incubated at 42 °C for 48 h under microaerophilic conditions (5.0% O_2, 10% CO_2, 85% N_2) in a 2.5-liter microaerophilic jar.
4. The 370 ml BPW will be transferred into a sterilized bottle and incubated at 35 °C for 24 h.

After 24 h

1. Subculture 0.1 ml of primary enriched BPW into 10-ml RV and 1.0 ml of enriched BPW into 10-ml TT medium, and incubate at 42 °C (RV) and 35 °C (TT) for 24 h.

After 48 h

1. A loop of secondary enriched RV and TT solution will be streak plated onto XLT-4 and HardyCHROM agar, and incubate at 35 °C for 24 h to 48 h.
2. A loop of Bolton broth will be streaked on modified Campy-Cefex agar and incubated at 42 °C for 72 h under aforementioned microaerophilic conditions.

Identification test

1. Presumptive *Salmonella* colonies (red colony with black center on XLT-4 agar and purple colonies on HardyCHROM agar) will be verified by Gram staining, *Salmonella* latex agglutination test, and API 20E test.
2. Presumptive *Campylobacter* colonies on the modified Campy-Cefex agar will be confirmed using the Campy-latex agglutination test and Gram staining (Gram-negative rods) to observe for corkscrew morphology.

Since We Already Practiced Gram Staining and Latex Agglutination Test, Please Write Major Steps of Gram Staining and Latex Agglutination Test Below

Gram Staining

Latex Agglutination Test

API 20E Test Procedure

1. Pick one to three presumptive *Salmonella* colonies from XLT-4 and HardyCHROM agar and suspend into a 5 ml of sterilized saline solution.
2. Use a bulb dropper to add the saline suspension into the cupules of the strip.
3. Fill the cupule with mineral oil for underlined test *ADH, LDC,* and *URE.*
4. Fill both the tube and cupule for *boxed tests CIT, VP,* and *GEL.*
5. Add 5 ml of water into the plastic base and place strip in the base.
6. Incubate the strip for 24 h at 35 °C.

After 24 h

1. IND: Add one drop of Kovac reagent to record results in 2 min.
2. TDA: Add one drop of 10% ferric chloride immediately.
3. VP: Add one drop of 40% KOH and then one drop of 6% alpha-naphthol in 10 min.
4. Record the results (+ or -) on the worksheet.
5. Record the API number and add number up for the positive reaction, and search the code in the API reference book.
6. Record and confirm the identity of the isolation in *Salmonella* spp.
7. A possible *Salmonella* spp. code is 6704752

Figure of API 20 E Result Sheet

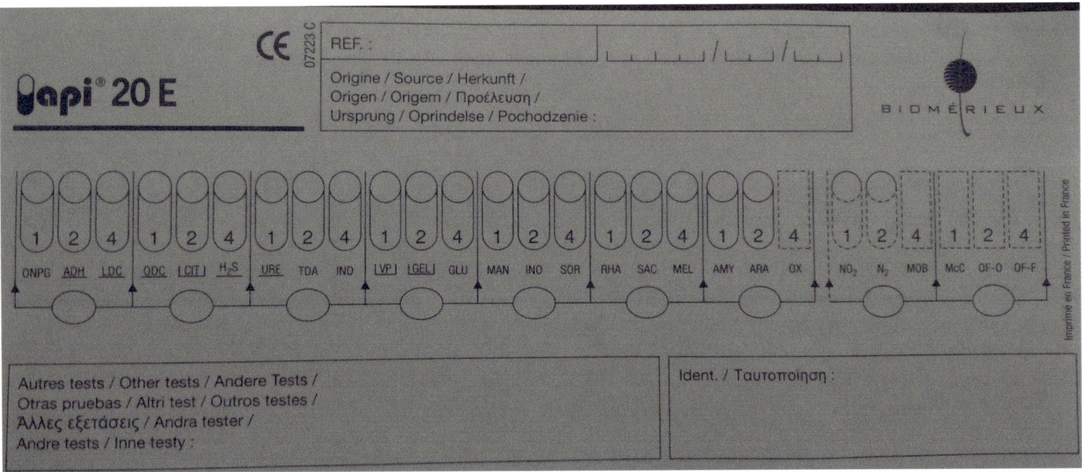

Review Question

Starting in July 2011, USDA-FSIS established new performance standards in response to national baseline studies requiring routine testing for *Salmonella* and *Campylobacter* in all processing plants. Questions: What are the performance standards? If you are a microbiologist working in a poultry company, what will you do for your company to meet this requirement, and if there are outbreaks happening, what will you do to isolate these two pathogens from contaminated chicken samples?

API 20 E test result table of color change:

Test name	Positive	Negative
ONPG	Yellow	Colorless
ADH	Red or orange	Yellow
LDC	Red or orange	Yellow
ODC	Red or orange	Yellow
CIT	Turquoise/dark blue	Light green/yellow
H_2S	Black deposit	No black deposit
URE	Red or orange	Yellow
TDA	Golden brown/red	Yellow
IND	Pink-red ring	Yellow
VP	Pink-red	Colorless
GEL	Diffusion	No diffusion
GLU	Yellow or gray	Blue to green
MAN, INO, SOR, RHA, SAC, MEL, AMY, ARA	Yellow-yellow green	Blue to blue green

Positive Salmonella spp. Result

Salmonella Typhimurium ATCC 14028 (positive control)

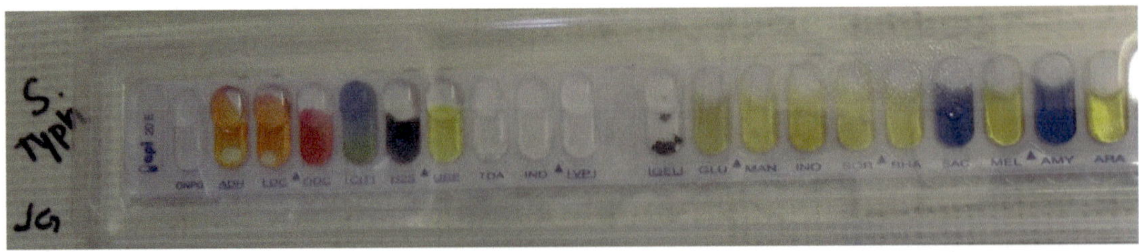

A *Salmonella* spp. positive sample isolated from chicken carcass

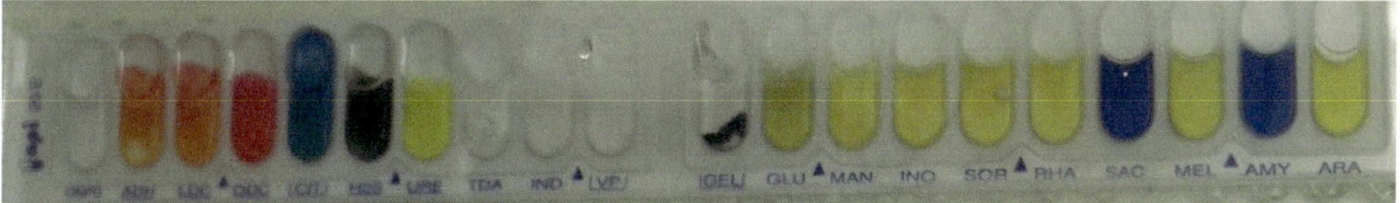

Lab work flowchart (*Salmonella*)

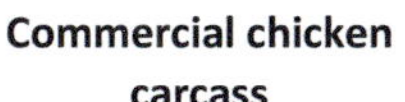

Commercial chicken carcass

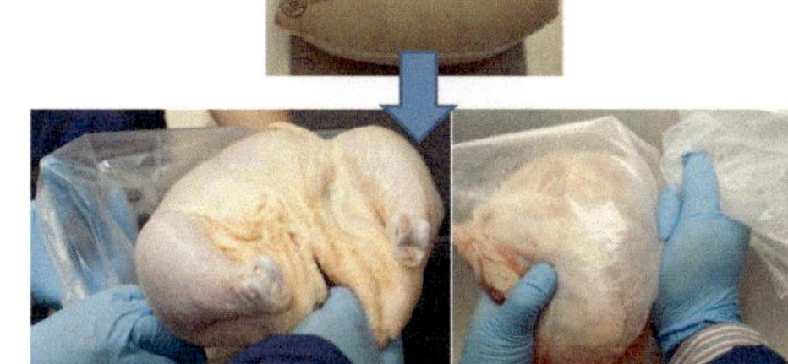

Transfer the carcass into poultry sampling bag with 400 ml BPW and shake 30 s

Carcass rinse water storage at 35°C for 24 h

(The residual chicken blood is good for enrichment)

Transfer 0.1 ml of enriched BPW into RV, incubate at 42°C for 24 h

Transfer 1.0 ml of enriched BPW into TT, incubate at 35°C for 24 h

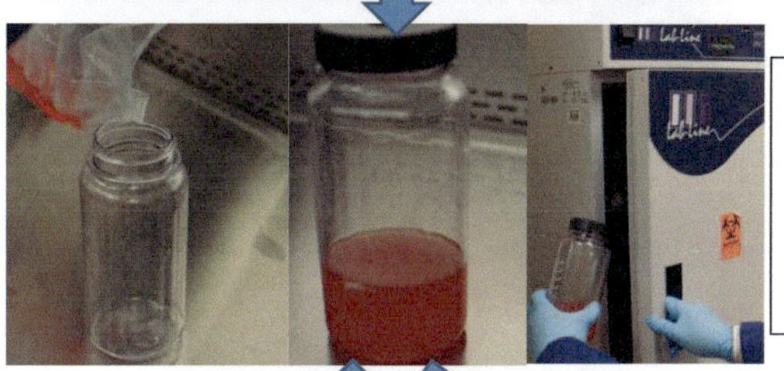

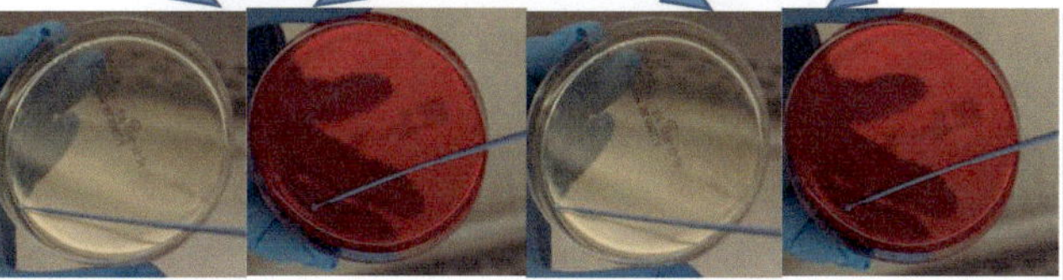

Streak-plating RV enrichment on HardyChrome and XLT-4 agar, incubate 35°C for 24h

Streak-plating TT enrichment on HardyChrome and XLT-4 agar, incubate 35°C for 24h

Lab work flowchart (*Campylobacter*)

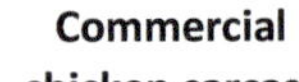

Commercial
chicken carcass

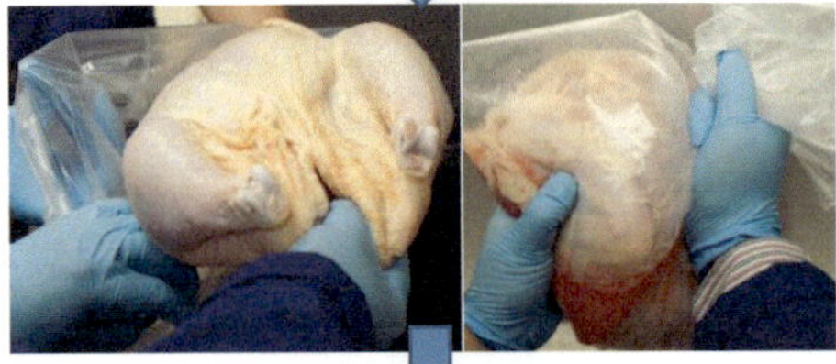

Transfer the carcass
into poultry sampling
bag with 400 ml BPW
and shake 30 s

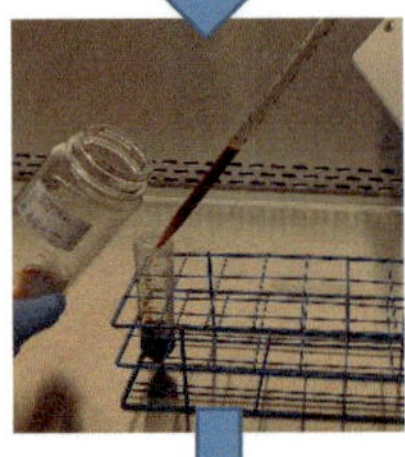

Transfer 25 ml of
shaken BPW solution
into 25 ml of Bolton
broth in a tube

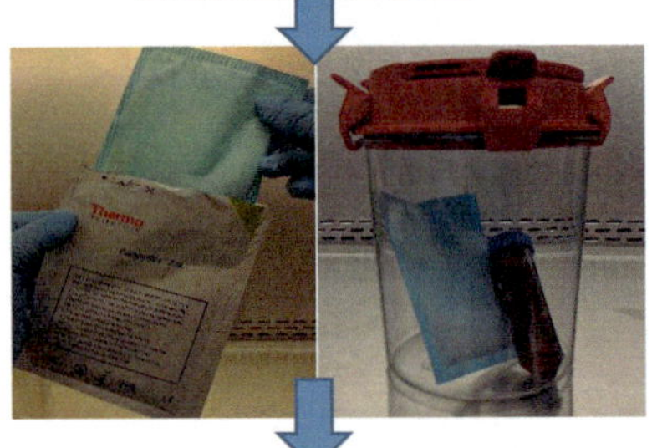

Store the tube into a
microaerophilic jar at
42.5°C for 48h.

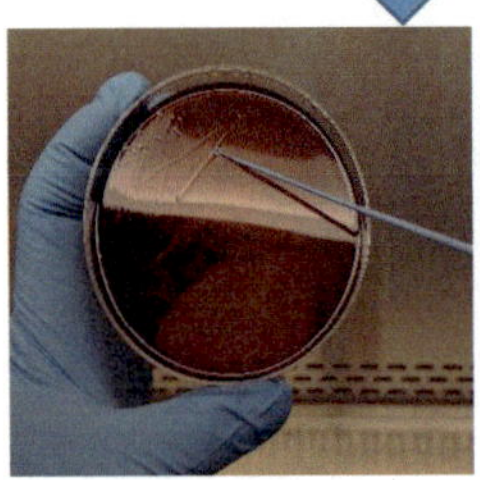
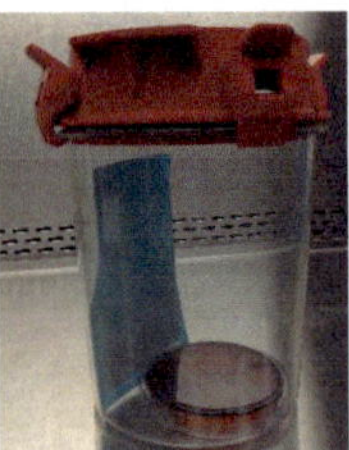

Streak-plating onto a
modified campy-cefex
agar and stored into a
microaerophilic jar at
42.5°C for 48h.

Part of Result Figures (Work from Previous Students)

A. Latex agglutination test

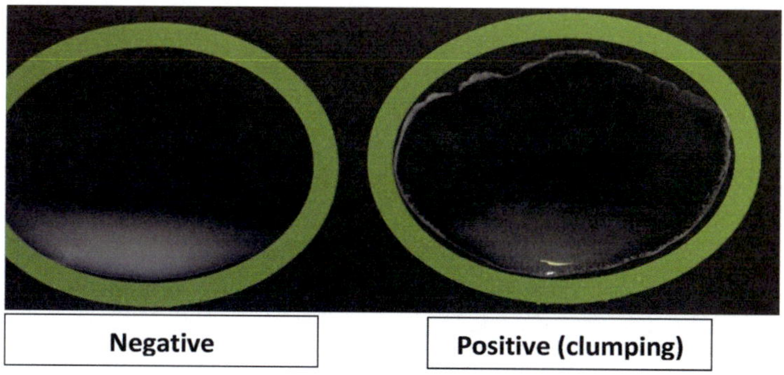

B. Gram staining results of presumptive Campylobacter spp.

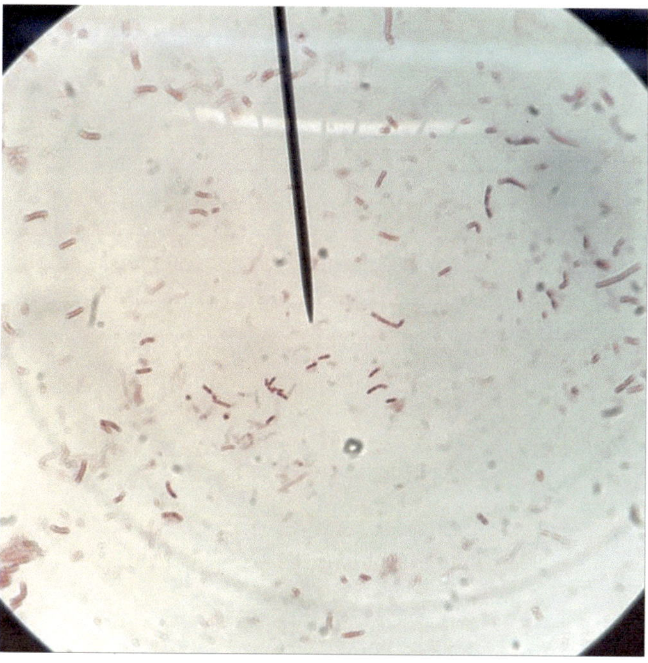

Gram-Negative, Single Rods with S Shape

Class Notes

Abstract

We will be manufacturing non-intact reconstructed beef patties, inoculating it with *E.coli* O157:H7 and aerobically storing in a PVC film-covered foam tray. We will evaluate the cooking inactivation activity of *E. coli* O157:H7 in beef patties with various cooking time. We will be enumerating survivals on support and selective medium and using USDA-IPMP 2013 software to analyze data and determine the fitted model. Finally, presumptive *E. coli* O157:H7 survivals will be identified using Enterotube II biochemistry test and latex agglutination test. In this chapter, it requires students to finish part of the work independently.

Keywords

E. coli O157:H7 • Non-intact beef • Thermal inactivation • Thermocouple • Latex agglutination test • MacConkey agar • Mug • Enterotube II test

Objectives Practice the cooking procedure to inactivate *E. coli* O157:H7 in manufactured non-intact reconstructed beef patties, and model the results using USDA-IPMP 2013 software.

Major Experimental Materials

- Fresh ground beef
- *E. coli* O157:H7 strains
- Benchtop meat grounder
- Kitchen aid and manual hamburger patty maker
- Foam trays and air-permeable plastic film
- Griller and type-K thermocouple
- Buffered peptone water (100% and 0.1%)
- Filtered food sampling bag
- MacConkey agar, MacConkey agar with MUG, and Tryptic Soy Agar
- *E. coli*PRO™ O157 Latex Kit
- Enterotube II

© Springer International Publishing AG 2017

C. Shen, Y. Zhang, *Food Microbiology Laboratory for the Food Science Student*,

DOI 10.1007/978-3-319-58371-6_9

Introduction As defined by the US Department of Agriculture's Food Safety and Inspection Service (USDA-FSIS), non-intact beef products include ground beef, mechanically or chemically tenderized beef cuts, restructured entrees, and meat products that have been injected with brining solutions for enhancement of flavor and/or tenderness. *Escherichia coli* O157:H7 (ECOH) or non-O157-Shiga toxin-generating *Escherichia coli* (STEC) can generate Shiga toxin and cause severe hemolytic uremic syndrome, with as little as ten cells causing death in a formerly health person. Since 1999, ECOH has been considered an adulterant of raw, non-intact beef products. On November 2011, the USDA-FSIS announced that, as of June 2012, non-intact beef products would also be considered adulterated if they were contaminated with non-O157-STEC serogroups O26, O45, O103, O111, O121, or O145.

Lab Work Procedure

Inoculation and Beef Patty Preparation: The meat will be manually cut into trimmings and then coarse grounded in a meat grinder (Gander Mountain #5 Electric Meat Grinder, Saint Paul, MN). Ground meat will be then mixed with 40 mL of the *E. coli* O157:H7 inoculum cocktail in a bowl-lift stand mixer (Kitchen Aid Professional 600, Benton Harbor, MI) at medium speed for 2 min to ensure even distribution of the inoculum into the sample, which simulates *E. coli* O157:H7 contamination during non-intact beef preparation. A manual hamburger patty maker (mainstays 6-ounce patty maker, Walmart, Bentonville, AR) will be then used to make beef or veal patties with 170–180 g of grounded meat.

Packaging: The reconstructed beef patties will be packaged aerobically in foam trays and covered using air-permeable plastic film and stored at 4.0 °C for 5 days.

Cooking: After the 5-day storage, the beef steaks will be taken out from their packages, weighed and double panbroiled in a Farberware@ griller with setup temperature of 177 °C (or 350 °F) for 0, 0.5, 0.75, 1, 1.5, 2, 3, 4, 5, 7, and 10 min. A type-K thermocouple was attached to the geometric center of the patty to monitor the internal temperature throughout the cooking with using PicoLog, a real time data recording software.

Enumeration

1. Cooked beef steaks (need to be weighted) will be sterilized and transferred into filtered-whirl food sampling bag immediately.
2. Add 200 ml of 0.1% buffered peptone water into the sample bag.
3. Stomach for 1 min.
4. Conduct serial dilution into 9.0 or 9.9 ml 0.1% BPW solution to obtain final dilution factor 10^{-1} and 10^{-3}, and spread plating onto MacConkey agar and Tryptic Soy Agar. Please draw your dilution diagram in your notebook.
5. Incubate plates at 35 °C for 48 h.
6. Manually count colonies of agar plates and record onto your notebook; calculate as follows:

Log_{10} CFU/g = Log_{10} [(Final CFU on plates/dilution factor)*(beef steaks gram + 200 ml)/beef steaks gram]

For example, after cooking, the beef is 100 gram, 50 CFU on plates with dilution factor 10^{-1}, and then final Log_{10}CFU/g

$= Log_{10} (50 \times 10^{1} \times 3) = Log_{10}1,500 = 3.2\ log_{10}CFU/g.$

Please draw your dilution procedure below:

Lab work flowchart

Course ground fresh beef

Inoculate *E. coli* O157:H7 in a kitchen kid

Making beef patties using manual patty maker

PVC film, aerobic storage for 48 h at 4°C

Double pan-broiling on a griller set at 176°C

Identification of *E. coli* O157:H7

Pick a typical presumptive *E. coli* O157:H7 from MacConkey agar (pink colony due to lactose fermentation positive) to conduct the following test:

1. Streak-plate onto *sorbitol MacConkey agar with MUG* and incubate at 35 °C for 24 h.

    Note: MUG is 4-methylumbel-liferyl-β-D-glucuronide. *E. coli* O157:H7 cannot ferment sorbitol, and it does not have enzyme β-D-glucuronidase; therefore, it is nonfluorescent white colony under UV light.

2. *Latex agglutination test (E. coliPRO™O157 Latex Kit)*
    (a) Place one drop of saline solution onto two wells of the test card.
    (b) Using sterilized loop, pick presumptive colony from MacConkey agar and mix with the saline solution.
    (c) *Shake (very important!)* the latex reagent bottle (control and test reagent) and place one free failing drop into each well.
    (d) Use a new sterilized loop to mix the regent with colony very well.
    (e) Rock the card for 1 min to observe *the clumping.*

3. *Enterotube II test*
    Enterotube II (picture below) is a tube of 12 compartmentalized, conventional agar media that can be inoculated rapidly from a single isolated colony on an agar plate. The media provided indicate whether the organism ferments the carbohydrates such as glucose, lactose, adonitol, arabinose, sorbitol, and dulcitol; produces H_2S and/or indole; produces acetylmethylcarbinol; deaminates phenylalanine; splits urea; decarboxylates lysine and/or ornithine; and can use citrate when it is the sole source of carbon in the medium. The Enterotube II is an example of a rapid, multi-test system used in the identification of unknown oxidase-negative, Gram-negative, rod-shaped bacteria of the family *Enterobacteriaceae*. The *Enterobacteriaceae* is a family of bacteria normally present in the intestinal tract of humans and animals. All members of the family ferment glucose and are oxidase-negative, facultatively anaerobic, Gram-negative rods with simple nutritional requirements.

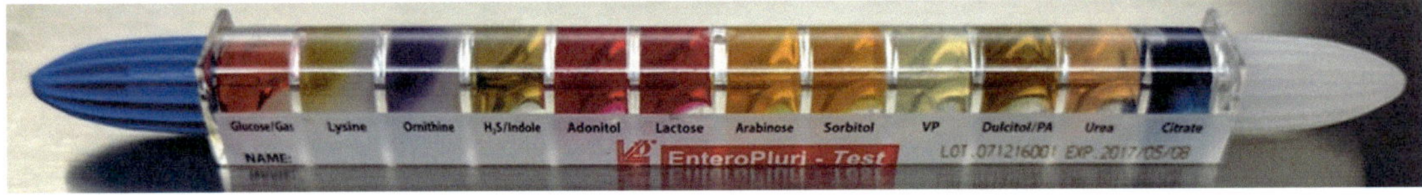

Procedure

Step 1. Remove the screw caps from both ends; aseptically pick one presumptive *E. coli* O157:H7 colony from MacConkey agar with the sharp end (white end).

Step 2. Twist the needle and pass it through the 12 wells and put it back to the original end, and screw the white cap back.

Step 3. Bend and break the needle at the blue cap end; break the "window" (punch the plastic film) of ADO, LAC, ARB, SOR, VP, DUL/PA, URE, and CIT compartments. Screw the blue cap back.

Step 4. Incubate the Enterotube II at 35 °C and read results in 24 h.

Step 5. Interpret the results of your Enterotube II using the instructions below.

Interpreting the Compartments: Circle the numerical value of each positive reaction.

Compartment 1. Interpret the results of glucose fermentation:
any yellow = +; red or orange = −
If positive, circle the number 2 under glucose on your Results pad.
Compartment 1. Interpret the results of gas production:
wax lifted from agar = +; wax not lifted from agar = −
If positive, circle the number 1 under gas on your Results pad.
Compartment 2. Interpret the results of lysine decarboxylase:
any purple = +; yellow = −
If positive, circle the number 4 under lysine on your Results pad.
Compartment 3. Interpret the results of ornithine decarboxylase.
any purple = +; yellow = −
If positive, circle the number 2 under ornithine on your Results pad.
Compartment 4. Interpret the results of H2S production:
true black = +; beige = −
If positive, circle the number 1 under H2S on your Results pad.
Compartment 4. Indole production.
Your instructor will give you the Indole test results of your unknown.
Compartment 5. Interpret the results of adonitol fermentation.
Compartment 6. Interpret the results of lactose fermentation.
Compartment 7. Interpret the results of arabinose fermentation.
Compartment 8. Interpret the results of sorbitol fermentation:
any yellow = +; red or orange = −
If positive, circle the points granted for each carbohydrate given on your Results pad.
Compartment 9. Voges-Proskauer test.
This test is not used unless a final VP confirming test is later called for.
Compartment 10. Interpret the results of dulcitol fermentation:
yellow = +; any other color = −
If positive, circle the number 1 under dulcitol on your Results pad.
Compartment 10. Interpret the results of PA deaminase:
black or smoky gray = +; any other color = −
If positive, circle the number 4 under PA on your Results pad.
Compartment 11. Interpret the results of urea hydrolysis:
red or purple = +; beige = −
If positive, circle the number 2 under urea on your Results pad.
Compartment 12. Interpret the results of citrate utilization:
any blue = +; green = −
If positive, circle the number 1 under citrate on your Results pad.

Once you have completed your biochemical readout, be sure you:

- Add all the circled numbers in each bracketed section and enter the sum in the space provided below the arrow on your Results page. You now have a five-digit reference number.
- Locate the five-digit number in the interpretation guide booklet and find the best identification in the column entitled "ID Value."

  Results pad for calculating identification code number:
  You may use the Results pad above to keep in your laboratory notebook.
  E. coli O157:H7 code should be 75340.

USDA-Integrated-Predictive-Modeling-Program Software

IPMP 2013 is a new-generation predictive microbiology tool. It is designed to analyze experimental data commonly encountered in predictive microbiology and for the development of predictive models. In this chapter, we will use survival models to analyze the results.

Brief procedures:

Step 1. Download and install "USDA-IPMP 2013 software" into your computer:

https://www.ars.usda.gov/northeast-area/wyndmoor-pa/eastern-regional-research-center/docs/
    integrated-pathogen-modeling-program-ipmp-2013/

Then, click "*Link to download IPMP2013.*"

Step 2. Enter cooking time into "x-data" column, and enter microbial raw data into "y-data" column. Click "submit raw data"; all raw microbial data will transfer to "log CFU/unit."

Step 3. Select the "Survival Models," and click the radio button of "Linear Model," "Gompertz Model," "Weibull Model," and "Two/Three-Phase Linear Model." For "Linear Model," choose both "with" and "without" tails.

Step 4. Your results should be fit to one of the following curves:

1. Linear curve
2. Linear curve with tail
3. Reparametrized Gompertz survival model
4. Weibull model
5. Linear curve with shoulder, two-phase model
6. Linear and tail three-phase model

Step 5. Find your RMSE and AIC number of each model equation; the best model that fits your results should have the smallest value of RMSE and AIC.

References (After Class Reading)

Huang, L. 2013. USDA Integrated Pathogen Modeling Program (http://www.ars.usda.gov/Main/docs. htm?docid=23355). USDA Agricultural Research Service, Eastern Regional Research Center, Wyndmoor, PA.

Huang, L. 2014. IPMP 2013 – A comprehensive data analysis tool for predictive microbiology. International Journal of Food Microbiology, 171: 100–107.

Review Question

Cooking ground beef patties at 176 °C for 0–15 min, we get the following results of *Escherichia coli* survival population; please use USDA-Integrated-Predictive-Modeling-Program software to calculate all parameters and choose the best model that fits the thermal inactivation curves (CFU/g, it needs to transfer to $\log_{10}$CFU/g):

Time (min)	0	1	2	3	5	7	9	10	12	15
CFU/g	2,000,000	1,200,000	1,100,000	1,000,000	90,000	20,000	5000	4000	200	50

Class Notes

Abstract

We will isolate anaerobic bacteria from locally canned food using an anaerobic jar with gas generator. The presumptive *Clostridium perfringens* will be identified on various selective media together with Gram staining and endospore staining. The pH value of the canned food will be measured.

Keywords

Anaerobic bacteria • Canned food • *Clostridium perfringens* • pH • Cooked meat medium • Tryptose sulfite cycloserine agar • Thioglycollate fluid agar • Anaerobic jar • Gas generator

Objectives Understand the anaerobic bacteria safety concern in locally produced canned food, measuring pH, and cultivation and identification of possible *Clostridium perfringens* in canned food.

Major Experimental Materials

- Canned food samples
- pH meter
- Blood agar
- Cooked meat medium
- Tryptose sulfite cycloserine agar
- Thioglycollate fluid agar
- Semisolid motility test agar
- Anaerobic jar and gas generator
- Sterile bacterial loop and needle

C. Shen, Y. Zhang, *Food Microbiology Laboratory for the Food Science Student*,
DOI 10.1007/978-3-319-58371-6_10

Introduction

Canned food is usually referred as low-acid food, with pH ~4.6 and water activity ~0.86. State of West Virginia set up the only standard for local canned food is pH ≤4.0. *Clostridium botulinum*, *Clostridium tetani*, *Clostridium perfringens*, and *Clostridium difficile* are the four major *Clostridium* spp. that could cause endospore in improperly canned food.

Source of trouble: low-acid foods that were improperly canned.

Trouble Signs

- Clear liquids turn milky
- Cracked jars
- Loose or dented jars
- Swollen or dented cans
- "Off" odor

Prevention

- Exam all canned foods before cooking.
- Cook and reheat food thoroughly.
- Keep cooked food away from "dangerous zone" 40–140 °F.

Symptoms After Eating　Double vision, difficult speaking, difficult swallowing, difficult breathing, gas gangrene, and it can be fatal.

Cooked Meat Medium　A medium for the cultivation of anaerobes provides muscle protein in the heart tissue granules for the growth of anaerobes. The muscle tissue also provides reducing substances, particularly glutathione.

Tryptose Sulfite Cycloserine Agar　Containing peptone and yeast extract to support growth of *Clostridium* spp. Some H_2S-reducing bacteria like *Clostridium perfringens* will reduce the sulfite to sulfide and generate black colonies with the formation of ammonium ferric citrate.

Thioglycollate Fluid Agar　A nutritive medium with a reducing agent (sodium thioglycolate) which removes oxygen from the broth. A chemical indicator, methylene blue, is included in the broth. The greenish to blue color indicates the presence of oxygen.

Procedure

1. Measure and record pH of assigned original canned food (95% ethanol rinsing pH meter before measuring pH).
2. Incubate the canned food samples for 10 days at 37 °C.
3. Examine the exterior of the container for any noticeable "trouble sign."
4. Sanitize the lid of canned food samples with 200 ppm chlorine.

Procedure of cooked meat medium:

5. Inoculate two tubes of cooked meat medium, one with 2 g canned food samples and the other tube with 2 ml of canned food liquid.
     Note: Cooked meat medium needed to be in flowing steam for 20 min immediately before using.
6. Incubate tubes at 37 °C for 5 days and record your observation.
     Note: *Clostridium perfringens* grows with gas generation.

Procedure of detecting possible Clostridium perfringens:

7. Streak-plate 100 µl of incubated canned food liquid onto tryptose sulfite cycloserine agar and blood agar.
     Note: *Clostridium perfringens* show black colony on tryptose sulfite cycloserine agar and double beta-hemolytic zone on blood agar.
8. Incubate agars in an anaerobic jar at 35 °C for 48 h.
9. Inoculate typical colonies (black color) into thioglycollate fluid agar at 35 °C for 24 h, and record the growth observation.
     Note: *Clostridium perfringens* grows thoroughly in the thioglycollate fluid agar.
10. Stab needle inoculate typical colonies into the middle of semisolid motility agar.
     Note: *Clostridium perfringens* is not motile.
11. Conduct Gram staining and endospore staining using typical colonies.

Please Write Down Gram Staining and Endospore Staining Procedure Below

Gram Staining

Endospore Staining

12. Measure the pH of final canned food samples

Review Question

What is low-acid food? If there are flooding in your area, how will you treat your canned food in your plant? Please discuss about the possible pathogen related to canned food.

Pictures

Principles of Anaerobic Jar

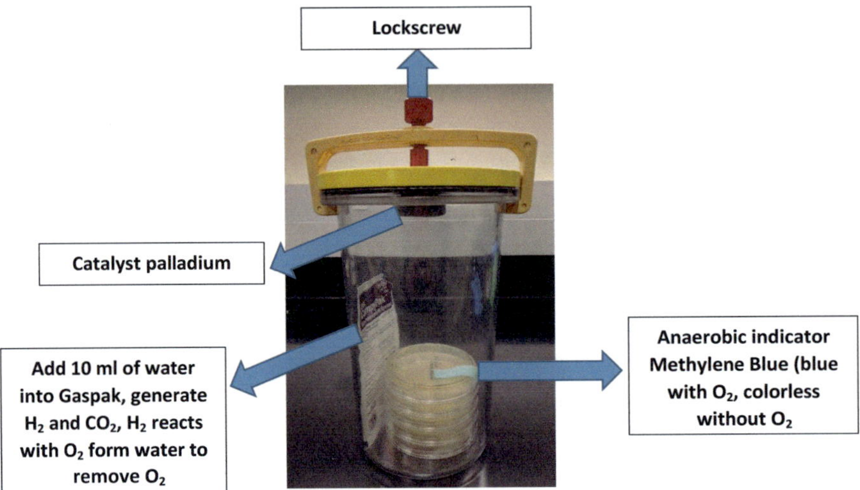

Result Figure

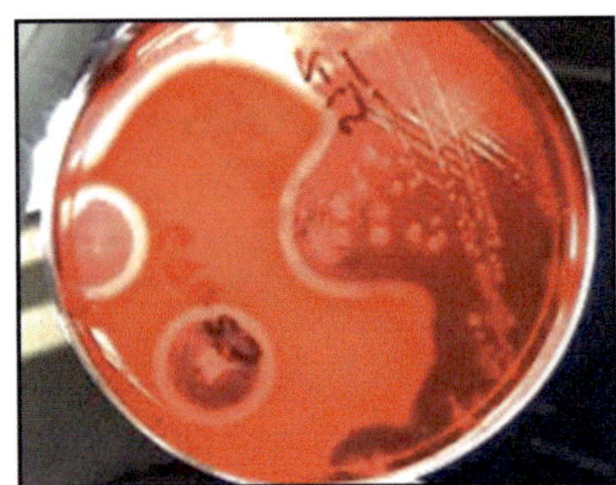

Clostridium perfringens cause double β-hemolytic zone on blood agar

Class Notes

____

____

____

____

____

____

____

____

____

____

____

____

Observation and Numeration of Molds from Spoiled Bread

11

Abstract

We will introduce the basic knowledge of molds from spoiled food. We will practice staining molds and observe their morphology. We will enumerate the population of fungi from spoiled bread using 3M@ yeast/molds Petrifilm and practice the dilution technique.

Keywords

Molds • *Aspergillus* • *Rhizopus* • *Penicillium* • Lactophenol cotton blue • Spoiled food • Petrifilm

Objectives Practice the lactophenol cotton blue staining to observe molds from spoiled bread and enumerate the population of molds using yeast/molds Petrifilm.

Major Experimental Materials

- Spoiled bread
- Lactophenol cotton blue
- Plastic tape
- Buffered peptone water (100% and 0.1%)
- 3M@ yeast/molds Petrifilm

Introduction of Fungi Fungi have dimorphic status, the yeast phase grows best at 35–37 °C, and their mold phase grows at 25 °C. The three popular mold types are *Aspergillus*, *Rhizopus*, and *Penicillium*.

© Springer International Publishing AG 2017
C. Shen, Y. Zhang, *Food Microbiology Laboratory for the Food Science Student*,
DOI 10.1007/978-3-319-58371-6_11

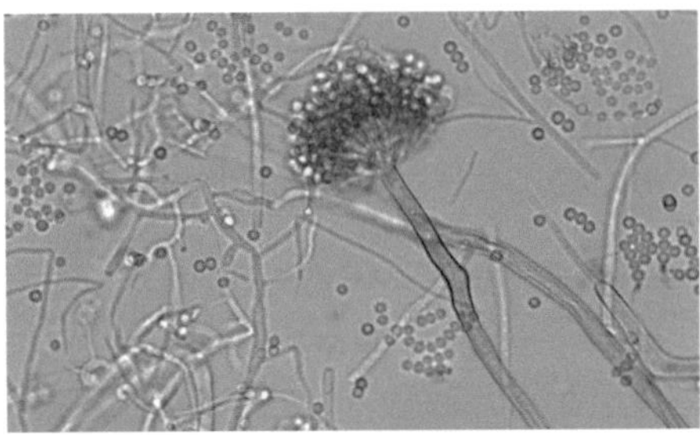

Aspergillus 100X – breads, grains, vegetables, dairy

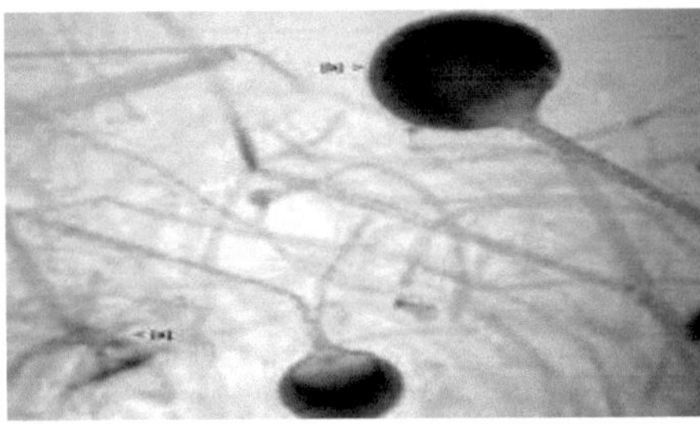

Rhizopus 100X–breads, vegetables

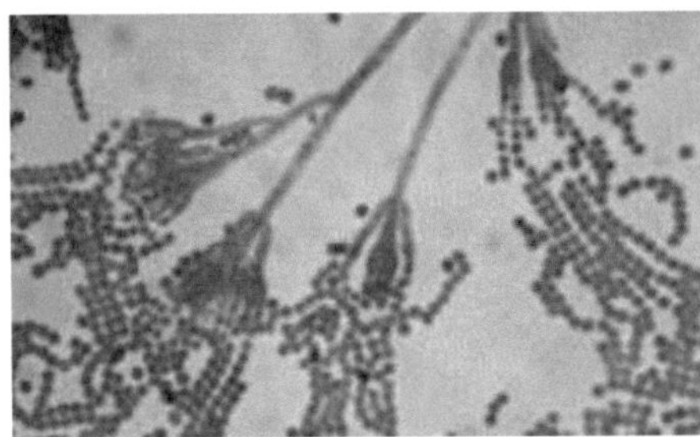

Penicillium 400X –breads, fruits, vegetables, dairy

Molds are composed of filament called *hyphae*, abundantly interwoven in a mat *(a very fuzzy appearance)* called *mycelium*. The mycelium extends upward from its vegetative base, thrusting specialized hyphae that bear *conidia* into the air, which is called *aerial hyphae.*

Procedure: Lactophenol Cotton Blue Stain

1. Obtain a mold-contaminated bread from the front desk.
2. Apply the tape to your index finger with the sticky side out, and do not allow the tape to stick to itself.
3. Open the bag of molded bread and gently apply the sticky side of the tape to the mold surface.
4. Place one drop of lactophenol cotton blue stain in the center of your clean microscope side.
5. Press the sticky side of the tape onto the drop of lactophenol cotton blue.
6. Observe the mold morphology under 10X and 40X objective lens.

Enumeration of Molds in Bread

1. Use chopstick to take out 15 g of spoiled bread and place into 50 ml buffered peptone water.
2. Stomach for 2 min, and spread plating onto yeast/molds Petrifilm with 10^{-3} and 10^{-5} dilution.
Note the presser used on yeast/molds Petrifilm is larger than the one used for APC, TCC, and *E. coli* Petrifilm.

Please Write Down Your Dilution Procedure Below

Review Question

Name and draw three possible molds in a spoiled bread, and briefly describe their structure.

Lab work flowchart

Molded bread

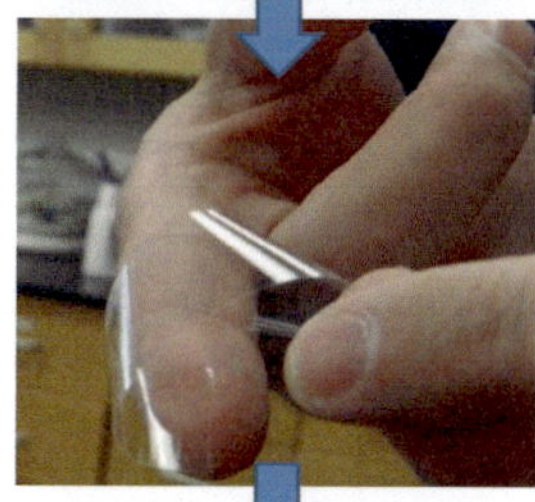

Apply the tape to your index finger with the sticky side out

Apply the sticky side of the tape to the mold surface

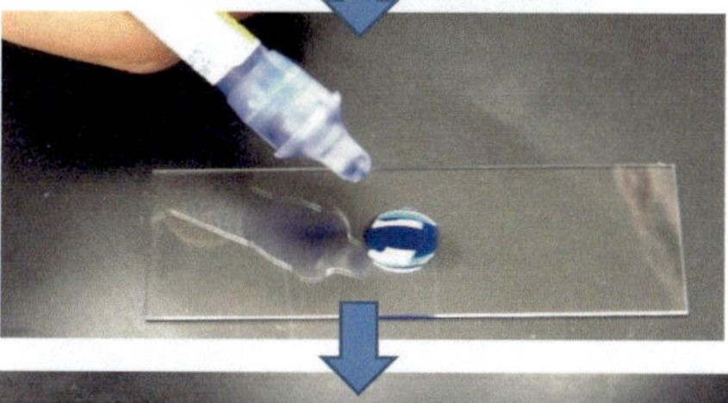

Add one drop of Lacto-phenol cotton blue on the center of glass slide

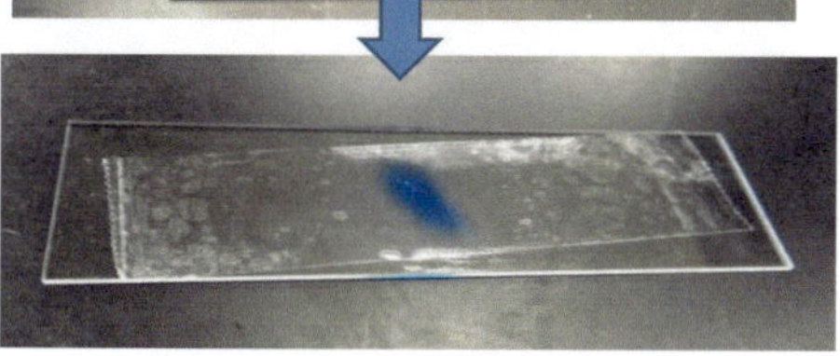

Press sticky tape onto glass slides, observe under 10X and 40X objective lens

Part of Result Figures
(Work from Previous Students)

A. Molds from spoiled bread grown on Petrifilm

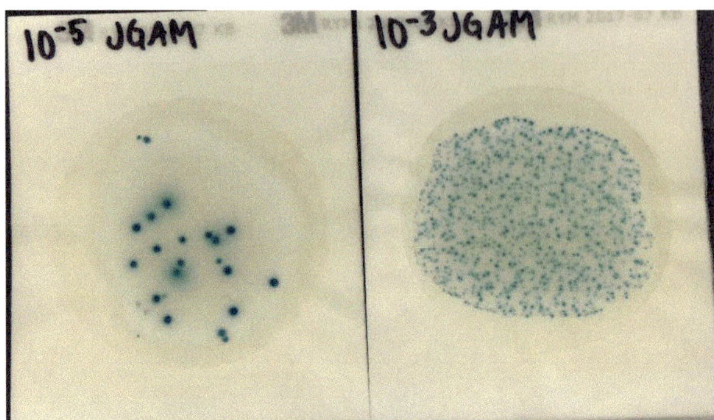

B. Molds from spoiled bread

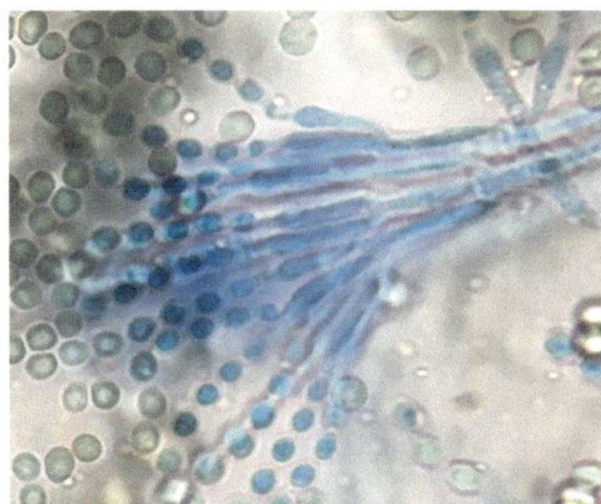

Penicillium (400X) from a spoiled bread

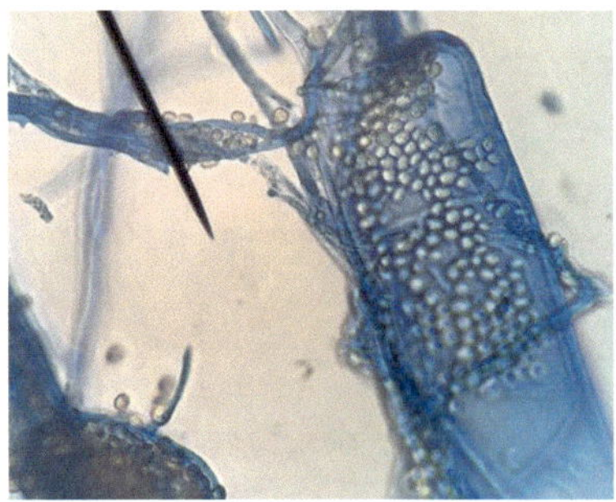

Possible Aspergillus (400X) from a spoiled bread

Class Notes

Abstract

We will introduce the PCR testing procedure for rapid identification of *L. monocytogenes* in deli meat. The lab work will include inoculation, extraction of DNA, PCR thermal cycle, and conduction of gel electrophoresis.

Keywords

PCR • *L. monocytogenes* • DNA • gel electrophoresis • Deli meat

Objectives Practice using PCR method to rapidly detect *L. monocytogenes* in deli meat (turkey breast or ham).

Introduction Polymerase chain reaction (PCR) is a DNA amplification method that produces many copies of a specific fragment of DNA. PCR can be used as a rapid method for bacteria identification in foods. Key ingredients in a PCR include forward and reverse primers that flank the DNA sequence to be amplified, a DNA template, *Taq* DNA polymerase, and dNTPs as building blocks for DNA synthesis. In PCR, DNA template is amplified using repeated cycles of denaturation (strand separation), annealing, and extension (strand synthesis). The target bacteria in this experiment are *Listeria monocytogenes* from a deli meat sample. The DNA template used for PCR will be extracted from the homogenized meat sample that has been inoculated with *Listeria monocytogenes* and followed an enrichment step to increase the number of target cells in relation to background microflora. After this experiment, students will become familiar with using rapid methods to detect foodborne pathogens and obtain a basic understanding of the concept and application of PCR.

Major Materials and Equipment

- Squeeze bottle filled with 70% ethanol or 10% bleach
- Gloves
- Deli meat contaminated with *Listeria monocytogenes*
- Graduated cylinder
- 250-ml flasks
- Buffered *Listeria* enrichment broth (LEB)

- DNA extraction kit
- Aerosol barrier pipette tips for PCR preparation
- Forward primer (P1) (10 pmol/μl)
- Reverse primer (P2) (10 pmol/μl)
- GoTaq Green Mix (containing PCR buffer, *Taq* DNA polymerase, and dNTPs)
- Deionized water
- Agarose
- 1XTBE buffer
- Ethidium bromide (EtBr)
- 100-bp DNA ladder

Equipment

- Scale
- Stomacher
- Stomacher bags
- Shaking incubator
- Thermocycler
- Pipetman
- Gel electrophoresis apparatus
- Gel documentation system

Procedure

1. Clean work area with 70% ethanol or 10% bleach.
2. Weigh 25 g of deli meat samples (pre-inoculated with *Listeria*).
3. Transfer sample to a stomacher bag containing 225 ml of buffered *Listeria* enrichment broth (LEB). Stomach for 2 min.
4. Transfer 50 ml of homogenized meat rinse to a 250-ml flask. Incubate at 30C with shaking at 100 rpm for 4 h.
5. Extract DNA using the DNA extraction kit.
6. Turn on thermocycler and set up the parameters as shown below.

94C 10 min ⟶ 94C   30s ⎤ 30 cycles ⟶ 72C 7min ⟶ 4C

55C   60s ⎦

72C   60s

7. Set up the PCR (25 μl). Dispense the following reagents into a PCR tube. Be sure that all of the reagents are mixed well.

DNA	2 μl
P1 (10 pmol/ μl)	1.25 μl
P2 (10 pmol/ μl)	1.25 μl
2XGo-Taq	12.5 μl
DI Water	8 μl

8. Run PCR cycle (approximately 2 h).
9. Cast a 1% agarose gel using 1XTBE buffer with 5 μl/100 ml of EtBr added.
10. Remove PCR tubes from the thermocycler. Load 15 μl of the PCR product to the agarose gel.
11. Run gel electrophoresis.
12. Visualize the PCR product in the gel doc system.
13. Clean work area with 70% ethanol or 10% bleach.

Lab work flowchart

Experimental Procedure of *Listeria* Identification by PCR

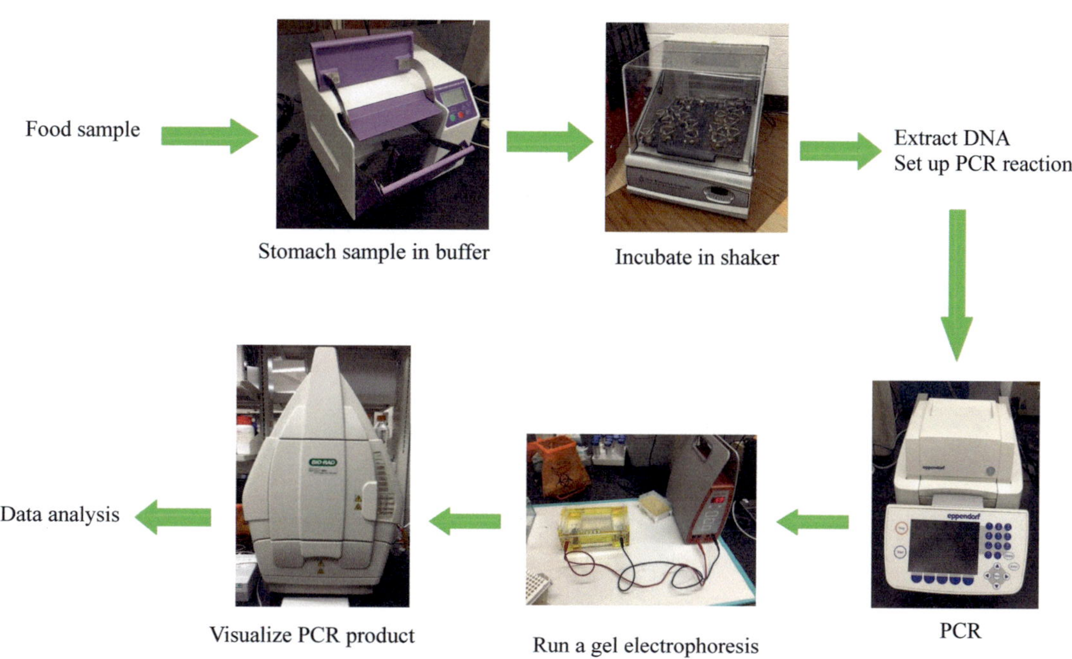

Class Notes

Abstract

We will practice manufacturing cheese using benchtop equipment with rennin and testing the quality parameters of cheese.

Keywords

Cheese • Rennin • Net weight amount • pH • Water activity

Objectives Practice benchtop cheese manufacturing process, and measure quality of made cheese including net weight amount, pH, and water activity.

Introduction The process of modern cheese making is a refinement of the techniques discovered thousands of years ago. Traditionally, fermentation of lactose (milk sugar) causes the milk to curdle due to a pH decrease, separating whey and curds. Another way of making cheese is by the addition of purified enzymes, such as rennin, a type of protease that cleaves the casein into small fragments that settle out as curds. Rennin works best at body temperature (37 °C). If the milk is too cold, the reaction is very slow, and if the milk is too hot, the heat will denature the rennin, rendering it inactive. There are many rennin cheeses, including Asiago, most brie, most cheddar, and Roquefort. Cheese making is dependent on a variety of factors, including the fat content of milk, pH of the solution, use of enzymes, curing time, and different treatments to influence taste, texture, and aroma.

Major Materials and Equipment Whole milk

- Rennin
- Hot plate
- 250-ml beaker
- Cheese cloth
- Thermometer
- Stirring rod
- Heatproof gloves
- Heatproof pad

© Springer International Publishing AG 2017
C. Shen, Y. Zhang, *Food Microbiology Laboratory for the Food Science Student*,
DOI 10.1007/978-3-319-58371-6_13

- pH meter
- Dehydrator
- Water activity meter

Procedure

Enzymatic coagulation of the casein from milk:

1. Weigh an empty beaker and record the weight.
2. Pour 250 ml of milk in the beaker. Weigh and record the weight
     (weight of milk = weight of beaker with milk − weight of beaker).
3. Set the hot plate at 43 °C (110 °F). Heat up the beaker with the milk on the hot plate.
4. Add three to four drops of rennin and two to three drops of vinegar, stir for 2 min, and allow the milk
     to sit on the lab bench for 5 min.
        Cover two to three layers of cheesecloth on a clean beaker. Pour the curds and liquid from the
     previous step into the clean beaker.
5. Gather up the cheesecloth and squeeze out the liquid whey from the curds. Spread out the cheese-
     cloth to allow the curds to dry for 5 min.
6. Weigh the curds (excluding the weight of the cheesecloth).
7. Calculate the yield of cheese by dividing the weight of curds by the weight of milk.

Variations:

8. Test the effect of low and high temperatures on the activity of rennin. Repeat the experiment with
     cold milk at 4 °C and hot milk heated to 70 °C.

pH measurement:

1. Rinse the electrode of the pH meter with distilled water.
2. Place 5 g of fragmented cheese in a test tube or small beaker.
3. Gently push the electrode into the cheese. Record the pH.

Measurement of the moisture content and water activity:

A. Determination of the moisture content
    1. Weigh approximately 100 g sample and record as "wet weight of sample."
    2. Set the dehydrator at no more than 239 ° F (115 ° C). Dry the sample in the dehydrator for 24 h.
    3. Allow the sample to cool.
    4. Weigh the cooled sample again, and record as "dry weight of sample."
    5. Calculate the moisture content using the following equation to the nearest tenth of 1%:

    $$\%W = A - B/B \times 100$$
    where W = percentage of water in the sample
        A = weight of wet sample
        B = weight of dry sample

B. Determination of the water activity
    1. Transfer a small amount of sample (to be able to cover the bottom of the sample cup) for water
       activity measurement.

2. Insert the sample cup into the water activity chamber.
3. Lock the chamber and read the water activity when the number is stable.

Lab work flowchart

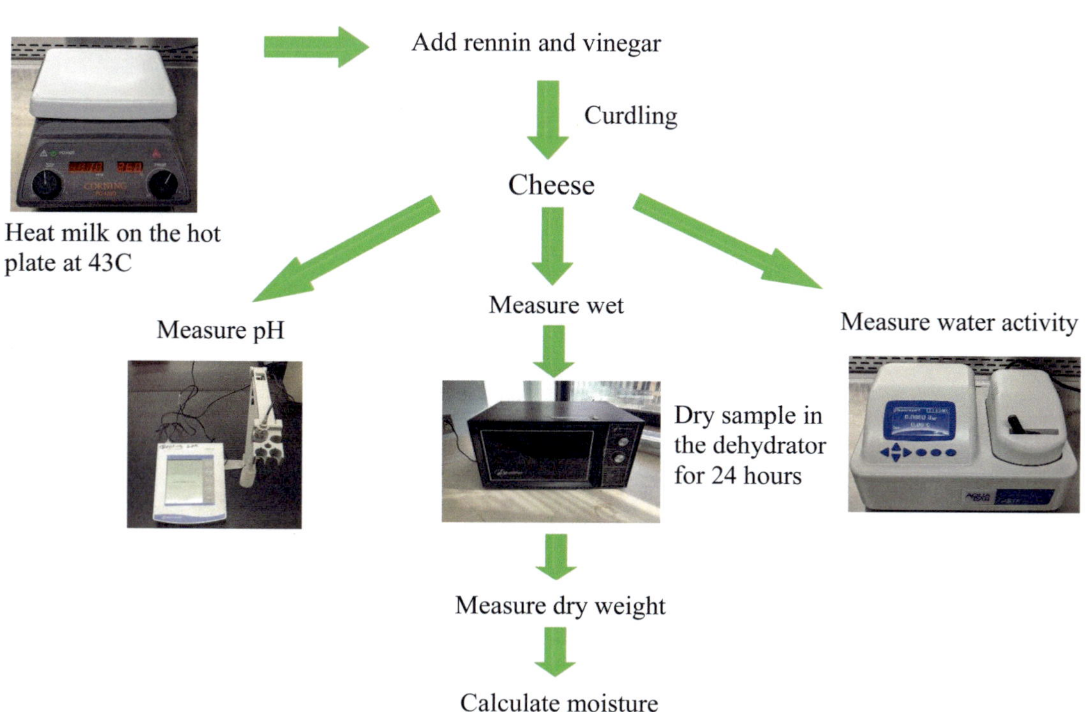

Class Notes

____
____
____
____
____
____
____
____
____
____
____
____
____
____
____
____
____
____
____
____

Abstract

We will introduce the biological function of bacteria and fungi during fermentation process and practice manufacturing wine and pickles using benchtop equipment and testing the quality parameters and microbial qualities.

Keywords

Wine • Pickles • pH • Water activity • Acid amount • Fermentation • Wine yeast • Microbial quality

Objectives Practice benchtop wine and pickle manufacturing process, and measure quality of made products including pH, water activity and acid amount, and microbial qualities.

Major Experimental Materials

- Lalvin QA23 white wine yeast
- 1-l and 500-ml flasks
- 1-l white grape juice
- A new balloon
- Sucrose
- Fresh cabbage
- 2.5% salt solution
- Fresh empty glass can with lid
- pH meter
- Water activity meter
- 1% phenolphthalein solution and 0.1 N sodium hydroxide solution
- 3M@ aerobic plate counts and yeast/molds Petrifilm and 0.1% buffered peptone water

Introduction Fermentation is an important step during food manufacturing history to preserve foods and includes homo-fermentation generating lactic acid and hetero-fermentation generating alcohol. Wine and pickles are two widely produced nondairy fermented foods. Wines are normal alcoholic

C. Shen, Y. Zhang, *Food Microbiology Laboratory for the Food Science Student*,
DOI 10.1007/978-3-319-58371-6_14

fermentation of grapes (containing sugar such as sucrose) with aging. The major difference between white and red wine is that white wines are fermented without the grape skins. Pickle is a fermentation product of fresh cucumbers with immersion into salt solutions (5%) for 6–9 weeks. Salt inhibits Gram-negative bacteria but supports lactic acid bacterial growth.

Procedure
Wine Preparation
Method 1 (grape juice)

1. Add 500 ml of commercial grape juice into a 1-l flask.
2. Add 10 g of sucrose and 0.6 g of Lalvin QA23 wine yeast.
3. Incubate the wine flask at 25 °C for 15 days.
4. Check the aroma (fruity, sweet, or none) and clarity (clear or turbidity).

Method 2 (grape)

1. Clean grapes with tap water.
2. Completely dry the grapes with paper towel.
   Note: Grapes must be completely dried out, otherwise molds will grow.
3. Use knife to cut grapes.
   Note: Please do not use hand, otherwise stem scar will be taken out, less yeasts will exist.
4. Weigh 300 g grapes into a clean jar and add 25 g sucrose.
5. Store the jar at 25 °C for 14 days.

Pickle Preparation

1. Clean cucumber with tap water and let it completely dry.
2. Use clean knife to shred cucumber (300 g) into fine pieces and add into a clean glass can.
3. Add 500 ml of the 5% salt solution into the glass can to overcover cucumbers about 0.5–1 in. The cucumber must always be immersed into the salt solution.
4. Incubate the cabbage jar at 25 °C for 42 days.

   Note: The above pickle preparation may end up with yeast/molds contamination, which will be used for the work of analyzing yeast/molds on Petrifilm. Please apply 100–200 ppm chlorine water to wash cucumber and add 5% vinegar if you want to avoid fungal contamination.

For Both Wine and Pickle, the Following Quality Items Will Be Tested:

1. Test the amount of *tartaric acid* (%): Take a 10 ml of the fermented wine or pickle solutions; mix with 10 ml of distilled water and five drops of 1% phenolphthalein solution. Titrate the amount of 0.1 N sodium hydroxide (NaOH) with first present pick color.
   % tartaric acid = (amount of 0.1 N NaOH × 0.1 × 7.5)/weight of samples (it is 10 ml)

2. Test *volatile acidity* (% acetic acid).
   % acetic acid = (amount of 0.1 N NaOH × 0.1 × 6.0)/weight of samples (it is 10 ml)
3. Test *pH*: Take a 6 ml of the fermented wine or pickle solutions to test pH using pH meter.
4. Test *water activity*: Take a 6 ml of the fermented wine or sauerkraut solutions using digital water activity meter.
   (a) Transfer a small amount of sample (to be able to cover the bottom of the sample cup) for water activity measurement.
   (b) Insert the sample cup into the water activity chamber.
   (c) Lock the chamber and read the water activity when the number is stable.
5. Take 10 ml of wine solution or 25 g of fermented pickles into 225 ml of 0.1% buffered peptone water and conduct tenfold serial dilution to test total *aerobic plate counts* and *yeast/molds counts* onto 3M@Petrifilm.

   Note: It is your assignment to conduct this test including establishing and drawing dilution procedure.
   Check with your instructor before you conduct the tests.

Draw Your Dilution Procedure Below

6. Fill the result table below:

Quality items	Wine 1	Wine 2	Pickles
% Tartaric acid			
% Acetic acid			
pH value			
Water activity value			
Aerobic plate counts			
Yeast/molds counts			

Note: We can use ebulliometer (costs ~$800–2500) to test alcohol amount in the wine (optional).

Lab work flowchart

Wine-1 (grape juice)

Fresh grape juice, wine yeast, and an autoclaved jar

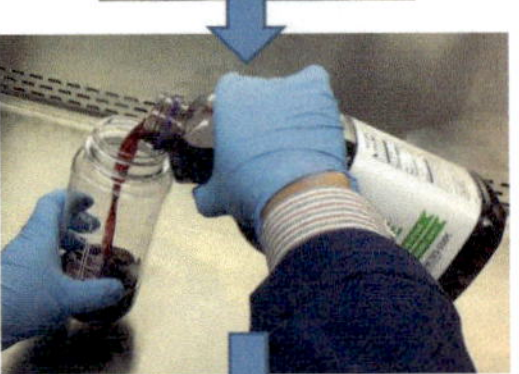

Add 500 ml of grape juice into the jar

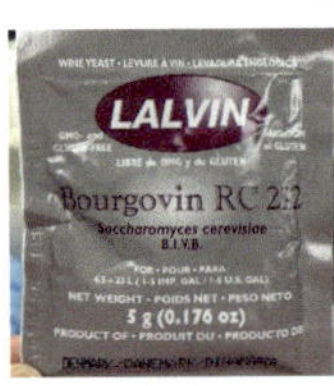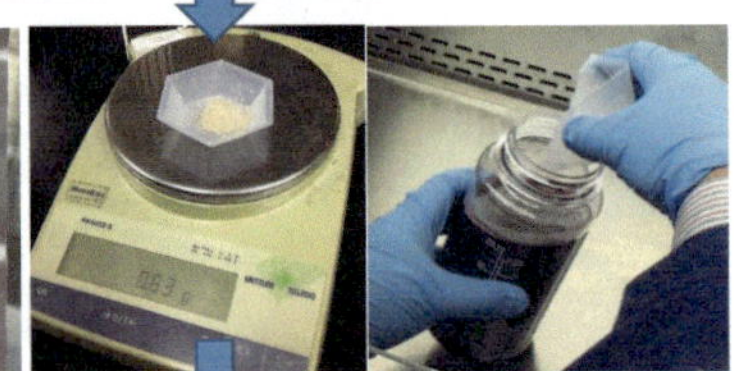

Add 0.6 gram wine yeast into 500 ml grape juice

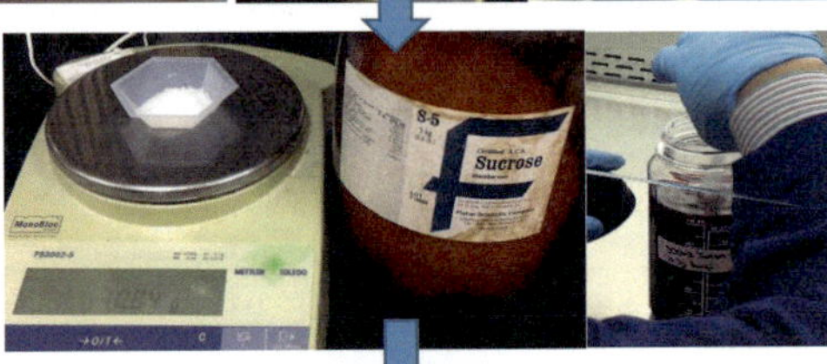

Add 10 gram sucrose into 500 ml grape juice

Incubate at 25°C for 14 days.

Lab work flowchart

Wine-2 (grape)

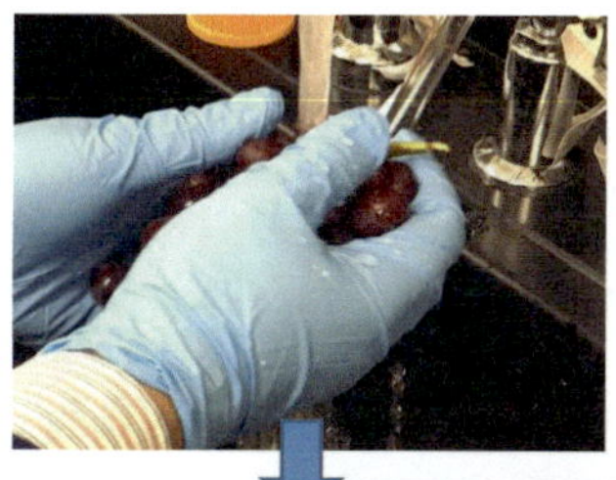

Wash grape under running tap water

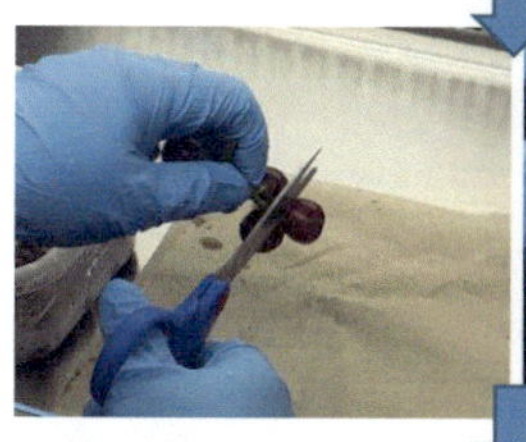

Cut grape using knife, if use hand stem-scar can be broken

Dry grape completely and add into jar

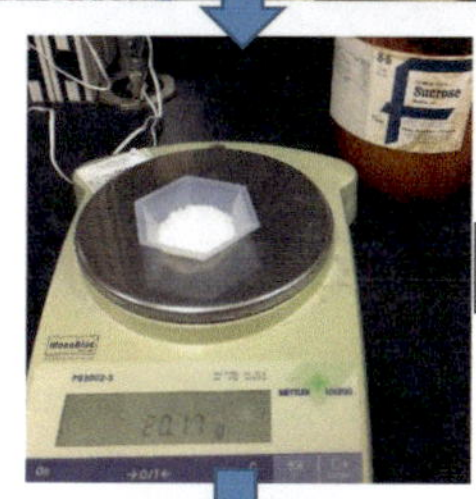

Weigh 20 gram of sucrose

Incubate at 25°C for 14 days.

Lab work flowchart

Pickles

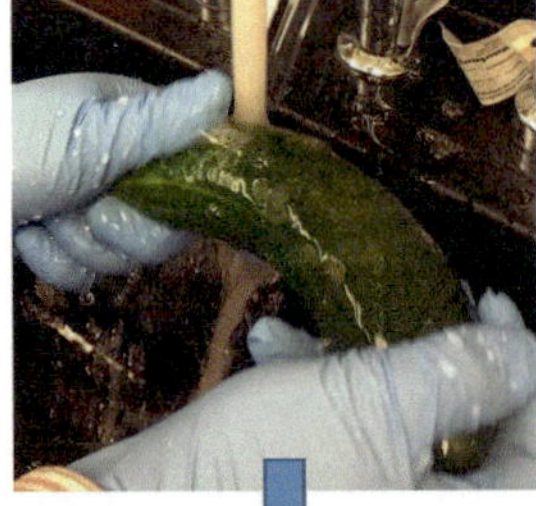

Wash cucumber under running tap water

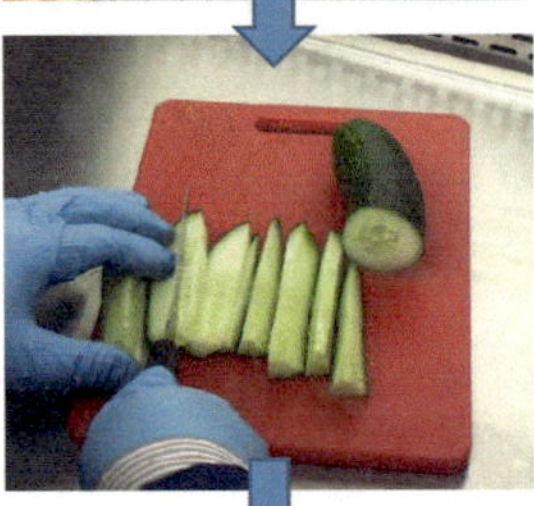

Trimming cucumber into small pieces

Add cucumber pieces into jar with 500 ml 5% salt solution

Saline solution need to be up cucumber pieces 0.5–1.0 inch, incubate at 25°C for 42 days

Class Notes

Antimicrobial Resistance of Commensal Bacteria from the Environment

15

Abstract

We will collect environmental samples and apply Kirby-Bauer test to evaluate the antimicrobial resistance of commensal microorganism.

Keywords

Antimicrobial resistance • Kirby-Bauer test • Commensal microorganism

Objective Apply disk diffusion test (Kirby-Bauer test) to determine the antimicrobial susceptibility of commensal bacteria from the environment.

Major Experimental Materials

- Tryptic Soy Agar (TSA) with tetracycline (8 µg/ml) and ceftiofur (4 µg/ml)
- Mueller Hinton Agar (MHA)
- 0.85% saline and buffered peptone water
- Cotton swabs
- Culture tubes
- Stomacher and vortexer
- Turbidity meter
- BD BBL Sensi-Disc Antimicrobial Susceptibility Test Discs
- Disk dispenser
- Ruler
- Incubator

Introduction Food and agricultural environment can be an important source of antimicrobial-resistant bacteria. Determining the antimicrobial susceptibility profiles of foodborne bacteria is important to understand the extent of antimicrobial resistance associated with our food supply. Disk diffusion test (Kirby-Bauer test) has been used since the 1940s to determine the antimicrobial susceptibility of bacteria. The principle is applying disks containing a wide variety of antimicrobial agents to the surface of Mueller Hinton Agar plates that have been inoculated with pure cultures of bacterial isolates. Following incubation, the plates are examined, and the zones of inhibition surrounding the

© Springer International Publishing AG 2017

C. Shen, Y. Zhang, *Food Microbiology Laboratory for the Food Science Student*,

DOI 10.1007/978-3-319-58371-6_15

"

disks are measured and compared with established zone size ranges for individual antimicrobial agents in order to monitor bacterial resistance and determine the agent(s) most suitable for use in antimicrobial therapy.

Lab Work Procedure

1. Each student collects two environmental samples such as river water, soil, wildlife animal, cattle and cow fecal samples, manure, weeds, grasses, and plant debris in their living area.
2. Ten grams of each sample will be added into 90 ml of buffered peptone water and stomached for 2 min.
3. Dilute above sample in a 0.1% buffered peptone water with 10^{-1} and 10^{-3} and spread plated onto Tryptic Soy Agar (TSA) supplemented with tetracycline (8 µg/ml) and ceftiofur (4 µg/ml) (they will be prepared ahead from lab).
4. Incubate your plates at 35 °C for 48 h, and manually count colonies.

We will use colonies from the TSA plus tetracycline (8 µg/ml) and ceftiofur (4 µg/ml) agar to conduct the following lab work.

OD Determination and Inoculum Preparation (This Step Can Be Done Ahead of Time)
   To make $1–2 \times 10^8$ CFU/ml of inoculums (i.e., 0.5 McFarland standard):

1. Add 5 ml of 0.85% saline to each of five culture tubes. Label the tubes with blank, OD 0.05, OD 0.08, OD 0.1, and OD 0.12.
2. Take one colony from TSA plus tetracycline (8 µg/ml) and ceftiofur (4 µg/ml) with a sterilized cotton swab and put into a labeled tube. Adjust OD value of each tube to 0.05, 0.08, 0.1, and 0.12 using a turbidity meter. Label each tube with 1 or 10^0.
3. Prepare four tubes for serial dilution for each OD value and label with 10^{-2}, 10^{-4}, 10^{-5}, and 10^{-6}. Add 990 µl, 990 µl, 900 µl, and 900 µl of saline to each tube.
4. Vortex tube 10^0 and transfer 10 µl to tube 10^{-2}.
   Vortex and transfer 10 µl to tube 10^{-4}.
   Vortex and transfer 100 µl to tube 10^{-5}.
   Vortex and transfer 100 µl to tube 10^{-6}.
5. Take out 100 µl from the last three dilutions and spread on TSA in duplicate. Incubate overnight at 35 °C.
6. Choose the 10^{-5} dilution with 100–200 colonies on plate. This corresponds to $1–2 \times 10^8$ CFU/ml in tube 10^0. Record the OD value. This OD will be used for the inoculum for disk diffusion.

Disk Diffusion Test
We will test antimicrobial susceptible to the following eight antimicrobial agents: ampicillin, clindamycin, erythromycin, gentamicin, methicillin, tetracycline, vancomycin, and streptomycin:

1. Aseptically pick one colony from your TSA plus tetracycline (8 µg/ml) and ceftiofur (4 µg/ml) agar.
2. Transfer the culture to 5 ml of 0.85% saline and adjust the OD to the predetermined value. This corresponds to $1–2 \times 10^8$ CFU/ml of inoculum (i.e., 0.5 McFarland standard).
3. Dip a sterile cotton swab into the tube and rotate it firmly several times against the upper inside wall of the tube to express excess fluid.
4. Streak the entire agar surface of MHA three times, turning the plate 60° between streakings to obtain even inoculation.

5. Leave the lid ajar for 3–5 min, but no more than 15 min, to allow for any surface moisture to be absorbed before applying the antibiotic disks.
6. Apply the disks using the disk dispenser and make sure disks are at least 24 mm apart.
7. Place the plate's agar side up in the incubator and incubate at 35 °C for 16–18 h.
8. Measure the diameters of the zones of complete inhibition as determined by gross visual inspection. Measurements are to the nearest whole millimeter.
9. Interpret the zone diameter using the breakpoints (in millimeter) on the Zone Diameter Interpretive Chart provided by BBL Sensi-Disc.
10. Fill the result table:

Antimicrobial agent	Code/concentration	Diameter of inhibitory zone (mm)				Your colony
		R	I	MS	S	
Ampicillin	AM10-10 mg	≤13	–	14–16	≥17	
Clindamycin	CC2-2 mg	≤14	15–20	–	≥21	
Erythromycin	E15-15 mg	≤13	14–22	–	≥23	
Gentamicin	CN10-10 mg	≤12	13–14	–	≥15	
Methicillin	MET5-5 mg	≤9	10–13	–	≥14	
Tetracycline	TE30-30 mg	≤14	15–18	–	≥19	
Vancomycin	VA30-10 mg	≤9	10–11	–	≥12	
Streptomycin	S10-10 mg	≤14	15–20	–	≥21	

Class Notes

Introduction of Oral Presentation and Job Interview Preparation

16

Abstract

In this last chapter, we will do food microbiology literature review, practice oral presentation and mini-thesis writing, and practice mock job interview.

Keywords

Literature review • Oral presentation • Job interview

Objectives Practice oral presentation with food science literature review and practice mock job interview questions.

Introduction By the end of this semester, we will all get hands-on experience of food microbiology technique especially on pathogen isolation/identification from various food products and learn basic experience of manufacturing different fermented foods. Many of us will be interested to look for an internship or a job in real food science world. Oral communication including oral presentation and job interview question/answer is an important step during job search; therefore, we will practice these two skills in this chapter to conclude the whole semester.

Practice of Oral Presentation Every student will do a food microbiology literature search, read and combine 2–4 peer-reviewed publications, and give 15–20-min oral presentation with 15–20 slides followed with submission of 2–3 pages of summary. All students, teaching assistant, and the instructor will evaluate the presentation.

Major Presentation Topics

- Pathogen control during food processing
- Function of beneficial bacteria during fermented food preparation
- Antibiotic resistant of pathogens in foods from farm to table
- Rapid testing of pathogens in foods
- Hazard analysis and critical control points (HACCP), good agricultural practices (GAP), and Food Safety Modernization Act (FSMA)-related topics

© Springer International Publishing AG 2017
C. Shen, Y. Zhang, *Food Microbiology Laboratory for the Food Science Student*,
DOI 10.1007/978-3-319-58371-6_16

Presentation Bodies

- Introduction
- Objectives
- Material and methods
- Results
- Discussion
- Implication (take-home message)

Suggested Peer-Reviewed Journals

- Journal of Food Science
- Journal of Food Protection
- Food Microbiology
- International Journal of Food Microbiology
- Food Control
- Applied & Environmental Microbiology
- LWT-Food Science & Technology
- Meat Science
- Poultry Science
- Journal of Applied Poultry Research
- Journal of Diary Science
- Journal of Animal Science

Suggested Literature Search Website

- Pubmed https://www.ncbi.nlm.nih.gov/pubmed/
- Google Scholar http://scholar.google.com/
- Science direct http://www.sciencedirect.com/
- Your university e-library system

Tips for a Good Oral Presentation

- Slides should be full with contents, do not leave lots of space.
- Not too many words on slides, an appropriate combination of words and pictures.
- Do not read through the words and sentences on slides, should paraphrase it.
- Speak loudly and audience can hear you clearly, have eye contacts.
- Clearly explain concepts, terminologies, data analysis, and results to audience.
- Answer questions in detail and be specific to the questions.
- Do not speak over time.
- Practice at least three times.

Presenter/Leader: Evaluation Rubrics (150 PTS)

Evaluator_________________________________

Presenter/Leader: _____________________________

Criteria	Yes	No
Presentation skills		
Was the leader organized and prepared (does not need to reread article)?		
Was the leader clear and understandable?		
Was the leader professional (confident, fluent, good pace, etc.)?		
Was there good use of communication aids (clear slides, font size, etc.)?		
Total YES = **/4**		
Presentation elements		
Did the leader provide pertinent and accurate background?		
Did the leader identify the author's central hypothesis and aims?		
Did the leader explain any uncommon/unfamiliar methods and statistics?		
Did the leader identify the important finding in the article?		
Did the leader accurately summarize the conclusions?		
Did the leader show non-bias?		
Did the leader identify unanswered questions and future directions?		
Total YES = **/7**		
Discussion leadership		
Was the leader able to answer questions?		
Did the leader involve class participants?		
Did the leader keep the discussion focused on issues related to the presentation?		
Total YES = **/3**		

Strengths

Areas for Improvement

Outstanding – 91–100% (13–14) of items scored, "yes"
Very good – 81–90% (11–12) of items scored, "yes"
Good – 71–80% (10) of items scored, "yes"
Satisfactory – 61–70% (9) of items scored, "yes"

Practice of Job Interview

Three major types of jobs for a food microbiology major student:

- Academic job: postdoctor, faculty, or other research/teaching/extension positions in a university
- Industry job: R&D scientist, food technologist, food microbiologist, QA/QC manager
- Government job: research microbiologist/food technologist at USDA ARS, FDA, or CDC.

In this chapter, we will practice a job interview for an entry-level food microbiology job in a company.

Tips for Job Search

- Get industry-scale working experience as much as you can.
- Get the professional certifications such as HACCP, GAP, and FSMA training.
- Apply and obtain US Green Card if you are an international student.
- Be a student member of food science professional associations, attend annual conferences, and involve in networking mixer with food industry personnel:
    - Institute of Food Technologist (https://www.ift.org)
    - International Association of Food Protection (https://www.foodprotection.org)
    - American Meat Science Association (http://www.meatscience.org)
    - Poultry Science Association (http://www.poultryscience.org)
- Learn the company and the position responsivity as much as you can.
- Tell your potential employer that you are a hardworking person, and do not tell that you know everything.

Mock Interview Practice You and your bench partner will be a group – one student is an interviewee and the other student is a human resource manager from a food company – ask the following questions, and then you two switch your position. Your instructor will give feedback of your answers.

Example John just got his bachelor's degree of food science from West Virginia University and has summer internship experience in a food commercial testing company at Pittsburgh; now he is looking for an entry-level microbiologist job in a company.

Ten Interview Questions

- Can you briefly introduce yourself?
- Why are you interested in this position?
- Why do you think we should hire you?
- Do you have any microbial working experience?
- What is your weakness?
- What is your greatest achievement in your lifetime?
- How can you think you are a good teamwork member?
- If there is an accident happening in a BSL-2 microbial lab, what you should do?
- What is the salary you are looking for?
- Do you have any questions for us?

Class Notes

Sample of Food Microbiology Lab Course Syllabus

Food Microbiology Laboratory Syllabus Sample

FDST449/549, 3:30–4:20 pm Tuesday and Thursday, 50 min/section, 2 sections/week, 1 credit, Agricultural Sciences South 1011

Instructor:	Dr. Cangliang Shen; TA: KaWang Li and Lacey Lemonakis
Office:	2423 Agricultural Sciences Building
Telephone:	304-293-2691 (Office; please leave message)970-222-2975 (Cell phone; please leave message)
Email:	cashen@mail.wvu.edu
Office Hours:	Tuesday and Thursday 4:30–5:00 pm or by appointment
Course Hours:	Tuesday and Thursday 3:30–4:20 pm
Prerequisite:	FDST 445/545: Food Microbiology Lecture

Course Description

This course is designed to give students an understanding of the role of microorganisms in food processing and preservation; the relation of microorganisms to food spoilage, foodborne illness, and intoxication; general food processing and quality control; the role of microorganisms in health promotion; and federal food-processing regulations. The listed laboratory exercises are aimed to provide a hands-on opportunity for the student to practice and observe the principles of food microbiology. Students will familiarize themselves with the techniques used to research, regulate, prevent, and control the microorganisms found in food and understand the function of beneficial microorganisms during the food manufacturing process.

Methods of Instruction

Laboratory course.

Expected Learning Outcomes

- Understand the principles of microorganisms during various food-processing and preservation steps.
- Isolation, identification, and enumeration of the most common microorganisms found in specific food products.
- Recognize specific types of microbial spoilage during various food shelf life stages.

© Springer International Publishing AG 2017
C. Shen, Y. Zhang, *Food Microbiology Laboratory for the Food Science Student*,
DOI 10.1007/978-3-319-58371-6

- Analyze different foods for presence of hazardous microorganisms using traditional and modern food microbiology technology.
- Describe the situations where improper food handling and storage may lead to the spoilage or contamination of food.
- Identify desirable microorganisms and their effects in preservation and fermentation.

Outlines of Topics

Date		Laboratory Experiment
January	10–12	Chap. 1. Food Microbiology Laboratory Safety and Notebook Record
	17–19	Chap. 2. Staining Technology and Bright-Field Microscope Use
	24–26	Chap. 3. Enumeration of Bacteria in Broth Suspension by Spread and Pour Plating
February	1–2	Chap. 4. Isolation of Foodborne Pathogens on Selective, Differential, and Enriched Medium by Streak Plating
	7–9	Chap. 5. Enumeration of Aerobic Plate Counts, Coliforms, and *Escherichia coli* of Organic Fruit Juice on Petrifilm
	14–16	Chap. 6. Enumeration and Identification of *Staphylococcus aureus* in Chicken Salads
	21–23	Chap. 7. Enumeration and Identification of *Listeria monocytogenes* on Ready-to-Eat (RTE) Frankfurters
March	1–3	Chap. 8. Isolation and Identification of *Salmonella* and *Campylobacter* spp. on Broiler Carcasses
	14–16	Chap. 8. Isolation and Identification of *Salmonella* and *Campylobacter* spp. on Broiler Carcasses
	21–23	Chap. 9. Thermal Inactivation of *Escherichia coli* O157:H7 in Non-intact Reconstructed Beef Patties
	28–30	Chap. 10. Cultivation of Anaerobic Bacteria in Canned Food
April	4–6	Chap. 11. Observation and Enumeration of Molds from Spoiled Bread
	11–12	Flexible (Plant Trip with Dr. J)
	18–20	Students' Presentation

Required Equipment

Biohazard hood, incubator, microscope, refrigerator, bench top meat grinder, griller, stomacher, centrifuge.

Attendance Requirements

****Attendance** for this course **is required** and will be graded as 2 points/class time. You must contact me no later than the lecture and laboratory period immediately following the date of your absence in order to determine whether your absence will be excused without the loss of 2 points. Preferably, I would appreciate hearing from you **in advance** of the laboratory session that you expect to miss.

Missed labs **cannot be made up** at a later time. *Thus, it is your responsibility to check with your laboratory partners for notes/information that you missed.*

Grading

The total number of points possible from the course is 200 points and is determined as follows:

Attendance **50 points**, 2 points/lab section/week

Note: two lab sections each week; on the first week, two lab sections are not calculated for attendance points.

Laboratory notebook record **50 points**

Middle-term exam **30 points**

Final exam **30 points**

<u>Presentation, mock interview, and mini-thesis</u> **40 points**
<u>Total</u> **200 points**

(Please note that there is a **requirement for attendance**. Also, laboratory notebooks will be collected for evaluation or grading.)

Grading Scale

A: 90–100%, $\geq$ 180 points
B: 80–89%, 160–179 points
C: 70–79%, 140–159 points
D: 60–69%, 120–139 points
F: less than 60% < 120 points

Availability of Lecture Notes

Laboratory handouts (PowerPoints and Word text outlines) are available on e-Campus or will be delivered by the instructor.

General Education Curriculum Statement

This course has been approved for inclusion in West Virginia University's General Education Curriculum (Group C of Objective 2 (Basic Mathematical Skills and Scientific Inquiry) or Objective 4 (Contemporary Society)). A significant part of this class will focus on increasing your ability to understand issues confronting society, understand an interdependent world, and use quantitative and scientific knowledge effectively.

Physical Handicaps Statement

If you are a person with a disability and anticipate needing any type of accommodation in order to participate in this class, please advise me after the first class meeting and make appropriate arrangements with Disability Services. Contact information for the Office of Accessibility Services is:

The Division of Diversity, Equity and Inclusion is located on the second floor at 1085 Van Voorhis Road in the Suncrest Center.

Phone: 304-293-6700Online:
http://accessibilityservices.wvu.edu/
Email: Access2@mail.wvu.edu

Per university policy, please **do not** request accommodations directly from the professor or instructor without a letter of accommodation from the OFSDS.

Inclusivity Statement

The West Virginia University community is committed to creating and fostering a positive learning and working environment based on open communication, mutual respect, and inclusion. If you are a person with a disability and anticipate needing any type of accommodation in order to participate in this class, please advise me and make appropriate arrangements with the Office of Accessibility Services (293-6700). For more information on West Virginia University's diversity, equity, and inclusion initiatives, please see http://diversity.wvu.edu.

Cell Phones Usage

Cell phones, headsets, and pagers are not to be used at any time during laboratory or during examinations. If you have any of these devices, they must be turned off during class.

Academic Integrity Statement

The integrity of the classes offered by any academic institution solidifies the foundation of its mission and cannot be sacrificed to expediency, ignorance, or blatant fraud. Therefore, I enforce rigorous standards of academic integrity in all aspects and assignments of this course. For the detailed policy of West Virginia University regarding the definitions of acts considered to fall under academic dishonesty and possible ensuing sanctions, please see the Student Conduct Code at http://www.arc.wvu.edu/rightsc.html. Should you have any questions about possibly improper research citations or references, or any other activity that may be interpreted as an attempt at academic dishonesty, please see me before the assignment is due to discuss the matter.

Laboratory Safety Rules/Procedures

We will cover food microbiology laboratory safety rules in our first class. Students are required to pass the lab safety quiz. Basically, students are required to wear gloves and lab coats, and no any open-toe shoes are allowed. Students are also required to wear goggles (some sections) during the laboratory course period, wash their hands before and after each lab section, and report any lab accidents immediately to the instructor.

Adverse Weather Statement

In the event of inclement or threatening weather, everyone should use his or her best judgment regarding travel to and from campus. Safety should be the main concern. If you cannot get to class because of adverse weather conditions, you should contact me as soon as possible. Similarly, if I am unable to reach our class location, I will notify you of any cancellation or change as soon as possible (by X o'clock/X hours before class starts), using e-Campus and email, to prevent you from embarking on any unnecessary travel. If you cannot get to class because of weather conditions, I will make allowances relative to required attendance policies, as well as any scheduled tests, quizzes, or other assessments.

Index

© Springer International Publishing AG 2017
C. Shen, Y. Zhang, *Food Microbiology Laboratory for the Food Science Student*,
DOI 10.1007/978-3-319-58371-6

Made in the USA
Monee, IL
07 July 2026

56546132R00071